Earth 2050 Global Collapse

A climate science fiction Novella by

J. B. Seeby

I0530495

This is a work of fiction. The names, characters, organizations, places, events, and incidents are used fictitiously. Any resemblance to actual persons, living or dead, or actual events is purely coincidental.

CONTENTS

INTRODUCTION

The world had gone to hell. The world of 2050 was a grim reflection of humanity's failure to curb the runaway train of global warming. Following the breakdown of the international meetings to control climate change the Paris Agreement and then NATO crumbled in 2050, marking the dawn of a chaotic new epoch. The relentless march of global warming had spiraled out of control, fueled by the unchecked consumption of fossil fuels. The planet now teeters on the brink, mired in environmental devastation, political turmoil, and social collapse.

Ever since people began tracking the amount of carbon dioxide in the atmosphere in 1958 atop Mauna Loa Observatory Hawaii the gas has been steadily increasing from 315 ppm to 850 ppm in 2050. This increased concentration of carbon dioxide and other greenhouse gases (gases that trap heat in the atmosphere) such as methane, nitrous oxide, and water vapor caused the Earth's surface temperature to rise 6° Celsius, or 42.8° Fahrenheit. This added amount of heat was caused by the energy trapping effects of the molecular structure of carbon dioxide trapping infrared radiation and heating the Earth, the so-called greenhouse effect. Once the Earth crossed the 450 ppm threshold of carbon dioxide, sometime in the mid-to-late 2030s, global warming became a runaway positive feedback loop wherein everything humanity did to try and cool Earth only exacerbated the problem. This was learning thermodynamics and entropy the hard way: for every high-technology action humanity took to decrease Earth's surface temperature, a greater amount of heat trapping greenhouse gases was produced, which increased Earth's surface temperature even more. Even if humanity stopped all production and consumption of fossil fuels on Earth, the heat trapping gases would remain in the atmosphere for a thousand years. We were trapped unless we could find a way around the second law of thermodynamics' entropy law. Was that even possible?

In early 2028 solar radiation modification, or geoengineering, was tested in India but it was an abysmal failure. The most widely discussed forms of geoengineering were to increase the quantity of solar radiation reflected back into space, including surface albedo enhancement, marine cloud brightening (MCB), stratospheric aerosol injection (SAI), and space-based methods. In contrast, cirrus cloud thinning (CCT) involved the reduction of cirrus clouds to increase the amount of terrestrial radiation "lost" from the Earth system. All these methods were supposed to alter fluxes of both long- and short-wave radiation. One method proposed injecting material into the atmosphere such as sulfur dioxide. The end of geoengineering came when a major US university's geoengineering project was abandoned in late 2028 after India's test failed and an advisory committee released a report based on an environmental impact statement which condemned the process of altering the atmosphere in a potentially dangerous fashion.

As great nations once considered unassailable economic giants crumbled under the weight of their own unsustainable practices, a cascade of famine and conflict ensued. The global community, fractured and devoid of any meaningful alliances, became a breeding ground for hostility and despair. Every day, new conflicts flared, communities broke apart, and the specter of an uncertain future loomed larger.

And yet, amid the shadows of despair, a profound hope arose. Unbeknown to the billions who cast their eyes to the dim horizon, a group of international scientists, hidden away on the island of Crete, worked tirelessly on the edge of the unimaginable—reversing the second law of thermodynamics on a continual basis in order to cool the Earth to somewhere in the 400 ppm level of carbon dioxide concentration. This small band of physicists, driven by the urgency of their mission, unraveled the intricacies of a discovery that defied nature itself—a means to reverse the second law of thermodynamics and reduce overall entropy here on Earth.

Sequestered in what was once a NATO missile firing installation, these men and women pursued their dangerous gamble, risking exposure with every step. Within the ancient walls, they dared to imagine a future where the principles of energy and entropy might be rewritten—a world where the inexorable advance of disorder is not a foregone conclusion but a malleable destiny.

Their burgeoning hope, a spark in the enveloping darkness, is the last, best chance for the people of Earth. As nations falter and societies crumble, the prospect of reversing the tide of global warming lies in their hands. Though the road ahead is fraught with peril and the potential for catastrophe, it is also paved with the glimmer of a sustainable tomorrow.

With the forces of nature and history aligned against them, they know the weight of their task. Within their grasp hangs the delicate balance of world sustainability or its further undoing. As the narrative of our planet teeters between these two monumental paths, the actions of a few may yet shift the course of fate and alter our shared reality for the better.

CHAPTER 1

The year 2050 didn't arrive with a singular, cataclysmic bang, but rather with the sickening crunch of buckling institutions. It was the sound of resignation, the noise a worn-out support beam makes just before giving way entirely. Many had seen it coming, but confirmation still landed like a hard punch.

The first casualty was the already fragile Paris Agreement. While adherence had been patchy for years, America's withdrawal signaled the end. There were no lengthy withdrawal processes this time, just a stark declaration that American economic interests superseded any non-binding climate platitudes. Without the world's largest economy (still, barely) even pretending to participate, the agreement became functionally meaningless. Other nations, citing American intransigence, quickly followed suit, prioritizing short-term economic survival over long-term planetary health. The era of cooperative climate action officially flatlined.

Almost simultaneously, the North Atlantic Treaty Organization shuddered. NATO had always depended heavily on American military and financial backing, but more importantly, on a shared sense of purpose. With "America Unleashed" as the guiding principle, that purpose evaporated. Quietly at first, then more overtly, American assets were withdrawn, joint exercises canceled, and funding streams choked off.

European leaders, initially aghast, soon descended into squabbling. Old rivalries, suppressed for decades under the NATO umbrella, resurfaced. Who would lead? Who would pay? Who could be trusted? Without the American linchpin, the alliance simply lost its structural integrity. By the close of 2040, NATO wasn't formally dissolved by treaty, but it ceased to function. Military commands went dark, intelligence sharing dwindled to a trickle, and the collective defense guarantee became a historical footnote. Europe was on its own, and deeply divided.

The geopolitical shifts sent immediate shockwaves through the global economy. International trade, already strained by regional conflicts and creeping protectionism, and increased American tariffs became significantly more complex. Supply chains, intricately woven across continents based on decades of relative stability, began to fray. Shipping insurance rates skyrocketed as naval patrols under NATO command disappeared from key routes. Nations scrambled to secure resources domestically or through precarious bilateral deals. The era of global power and interdependence was rapidly crumbling.

Energy markets reacted with predictable volatility. While the root cause of the global warming was increased fossil fuel consumption, the sudden political instability ironically made securing those same fuels *more* difficult and expensive. Nations hoarding resources, disrupted transport lines, and the collapse of international energy agreements led to price spikes and shortages, even as the long-term consequences of burning the stuff became terrifyingly apparent. It was a vicious cycle: instability bred energy crises, which fueled further instability.

The environment, of course, paid no heed to political maneuvering. Freed from the already inadequate restraints of Paris, global emissions didn't just continue; they accelerated. The positive feedback loops climate scientists had warned about for decades began to bite hard. Melting Arctic permafrost released vast stores of methane, a potent greenhouse gas, driving temperatures higher still. The shrinking North polar ice cap meant the Arctic Ocean absorbed more solar radiation, further warming it.

The immediate, tangible effects became undeniable. The "hundred-year flood" became an annual event in many coastal cities. Bangladesh, the low- lying Netherlands, large parts of Florida – these weren't future projections anymore, they were scenes of recurring disaster. Seawalls, built at enormous expense just a decade or two earlier, were overwhelmed. Retreat from the coasts began not as organized policy, but as a desperate, chaotic scramble by those who could afford to move.

Agriculture was hit particularly hard. Shifting weather patterns turned former breadbaskets into dust bowls, while regions previously too cold for large-scale farming faced their own challenges with unpredictable rainfall and new pests. The carefully calibrated system of global food production, reliant on stable climates and predictable seasons, started to break down. Yields dropped significantly in major exporting nations like the US Midwest, Ukraine, and Australia.

The first whispers of famine weren't concerned with aggregate global supply, but with distribution and access. As national economies weakened and international trade seized up, countries reliant on food imports found themselves unable to secure or afford staples. Rationing began in poorer nations, then spread to middle-income ones. Images of empty supermarket shelves, once confined to history books about wartime, started appearing on news feeds from supposedly developed countries.

Social cohesion began to erode under the strain. Governments, weakened by economic woes and the collapse of international frameworks, struggled to maintain order or provide basic services. Trust in authority plummeted. People increasingly looked to local communities, or worse, to tribal or factional loyalties, for security. The "us versus them" mentality, always simmering beneath the surface, boiled over with alarming frequency.

Migration pressures intensified dramatically. It wasn't just about escaping poverty anymore; it was about escaping drought, flood, and starvation. Climate refugees, numbered in the millions, began moving across borders, seeking sanctuary in regions perceived as less affected. This inevitably led to friction. Nations already struggling to feed their own populations viewed the newcomers with suspicion and hostility. Border walls, both physical and bureaucratic, went up around the globe.

In the United States, the "America Unleashed" sentiment transformed into isolationism tinged with paranoia. Having withdrawn from global responsibilities, the nation focused inward, but found itself grappling with the same climate-driven disasters as everyone else. Massive wildfires in the West, intensified hurricanes battering the Gulf and Atlantic coasts, and the slow-motion disaster of the Ogallala Aquifer drying up put immense strain on state, local, and federal resources.

Europe, lacking the unifying structure of NATO, drifted towards regional blocs. Northern European nations, generally wealthier and initially less impacted by the most severe climate effects, tried to insulate themselves. Southern European countries, facing desertification, water shortages, and the brunt of migration from North Africa and the Middle East, grew increasingly desperate and resentful. Tensions flared along new lines drawn by climate vulnerability rather than past Cold War demarcations.

Asia saw a complex reshuffling of power. China, while also suffering environmental consequences, sought to leverage the global chaos to expand its influence, particularly as American power receded. However, its own internal stability was threatened by water scarcity in the north and coastal flooding in the south. India grappled with catastrophic monsoon failures and record heatwaves that killed millions and rendered parts of the country uninhabitable for months at a time. Regional conflicts simmered, occasionally boiling over into border wars fought over dwindling natural resources like arable land and water.

Africa, disproportionately affected by global warming despite contributing least to its causes, faced a maelstrom of drought, famine, and conflict. The Sahel region continued its relentless expansion southward, displacing millions. Water wars erupted along major river systems like the Nile and the Niger. International aid, already dwindling for years, dried up almost completely after the United States discontinued the U.S. Agency for International Development in 2028. Other donor nations focused on their own crises. Governance collapsed entirely in several states, creating vast ungoverned territories.

The collapse wasn't uniform. Pockets of relative stability remained, usually in regions with geographic advantages, strong (often authoritarian) local governance, or simply by sheer luck. But these became islands in a sea of growing chaos. Technology, once seen as a potential savior, proved largely ineffective against the scale of the unfolding disaster. Geoengineering schemes were discussed but deemed too risky or politically impossible to coordinate in the fractured global landscape. Carbon capture technologies were unproven and remained too slow and expensive to make a meaningful difference against accelerating GHG emissions.

People tried to adapt, of course. Coastal communities experimented with floating architecture. Farmers switched to drought-resistant crops or indoor vertical farming where resources allowed. Water conservation became a necessity. But these were localized coping mechanisms, not systemic solutions. They were band-aids on arterial wounds. The fundamental drivers – increased GHG emissions, collapsing cooperation, resource scarcity – continued unabated.

The psychological toll was immense. A pervasive sense of anxiety settled over the world. The future, once a source of optimism for many, now looked bleak and threatening. Rates of depression, suicide, and social unrest climbed steadily. Conspiracy theories flourished, blaming scapegoats ranging from corporations and specific nations to shadowy cabals, further eroding social trust and making collective action even harder. The sheer scale of the problem induced a kind of paralysis in many.

By the late 2040s, the world map was being redrawn not by armies, but by climate and desperation. Icecaps were shrinking, coastlines were retreating, deserts expanding, and fertile lands shrinking. Political borders remained, but their relevance was increasingly challenged by the movement of people and the breakdown of national authority. The initial collapse of the Paris Agreement and NATO triggered political detonations that had set in motion a cascade of failures that was gathering momentum.

The hopeful rhetoric of previous decades about green transitions and sustainable development now sounded like bitter mockery. The brief window for effective action had been missed, squandered through political inertia, short-sighted nationalism, and a collective failure to grasp the existential nature of global warming.

The world hadn't ended in 2050. But the intricate systems that sustained modern civilization had begun to irrevocably unravel. The ties that bound nations, economies, and societies were fraying, thread by thread. This was the beginning of the long descent, the initial phase where the cracks became impossible to ignore, and the chilling reality of a planet pushed beyond its physical and social limits started to sink in. The chaos of 2050 wasn't born overnight; its seeds were sown in the political failures and environmental neglect of the preceding decades, and they began to sprout with terrifying vigor in the aftermath of the elections of 2048. The collapse had truly begun.

CHAPTER 2

The year 2050, that grim marker etched into the collective memory of a broken world, was not the beginning of the end. It was merely the year the bottom finally fell out. The rot had set in decades earlier, spreading through the foundations of global order and environmental stability like a silent, pervasive cancer. The dramatic collapse of international cooperation and the constitutional crisis in America were not bolts from the blue; they were the culmination of long-simmering tensions, ignored warnings, and a cascade of tragically poor decisions made under the lengthening shadows of the past. Understanding the catastrophe of 2050 requires looking back beyond the immediate trigger, into the murky decades that preceded it.

The early twenty-first century was an era defined by a profound paradox. On one hand, scientific understanding of anthropogenic global warming reached an undeniable consensus. Data poured in from melting glaciers, rising seas, and increasing greenhouse gases with attendant rising surface temperatures. Complex climate models, refined year after year, painted remarkably consistent pictures of a dangerous future. The Intergovernmental Panel on Climate Change (IPCC) issued increasingly stark warnings, pleading for urgent, coordinated action to slash greenhouse gas emissions. The diagnosis was clear, the prognosis alarming.

Yet, simultaneously, this era was marked by bewildering inaction and deliberate obstruction. Powerful vested interests, primarily within the fossil fuel industry and its backers, mounted sophisticated and well-funded campaigns to sow doubt about the science, exaggerate the costs of transition, and lobby politicians to delay or block meaningful climate policies. They cultivated an environment where uncertainty and misinformation, however manufactured, became a justification for paralysis. Decades were lost arguing about the validity of facts that were already overwhelmingly established within the scientific community.

Political polarization turned global warming into a toxic partisan battleground. What should have been a matter of shared risk and scientific reality became entangled in ideological warfare. Proposals for carbon pricing, renewable energy mandates, or international agreements were often dead-on arrival, sacrificed on the altar of political point-scoring or short-term economic anxieties. Cooperation fractured not just between nations, but within them.

While America became one of the most visible symbols of eroding global cooperation, the trend was worldwide. A wave of nationalist sentiment washed over numerous countries in the 2030s and 2040s. Leaders promising to put their nations "first" gained traction, often defining "first" in narrow, zero-sum terms. Multilateralism fell out of fashion. International bodies like the United Nations saw their influence wane, hampered by Security Council vetoes reflecting great power rivalries and a general reluctance by member states to cede sovereignty or resources.

Trade disputes escalated into low-grade economic warfare, disrupting supply chains and fostering mistrust long before the 2050 collapse shattered them completely. Regional conflicts, often fought through proxies or fueled by resource competition exacerbated by early climate impacts, further strained international relations. Disinformation campaigns, expertly wielded by state and non-state actors, deliberately sought to undermine trust in international institutions, foreign governments, and even objective reality itself, making good-faith negotiation almost impossible.

The Paris Agreement, even before America formally withdrew, was already showing signs of terminal weakness. Its non-binding nature, celebrated initially as a diplomatic triumph necessary to achieve consensus, proved to be its Achilles' heel. Nationally Determined Contributions (NDCs), which outlined each country's efforts to reduce greenhouse gas (GHG) emissions, were often insufficient to meet the agreed-upon temperature goals, and mechanisms for enforcement or ratcheting up ambition proved inadequate in the face of entrenched domestic opposition and nationalist backsliding. It became a symbol of good intentions undermined by a lack of political will.

The environment, meanwhile, wasn't waiting for humanity to get its act together. The shadows of the past weren't just political; they were meteorological. The 2020s and 2030s saw a disturbing acceleration of extreme weather events that ought to have served as deafening alarms. There were intense hurricanes in 2028, which devastated Puerto Rico and lingered, dumping unprecedented rainfall across the Caribbean for nearly a week. Then came the heat domes of 2031, shattering temperature records from Portugal to Pakistan and causing widespread crop failures and millions of deaths.

Australia witnessed an endless summer turn into a recurring nightmare, with fire seasons growing longer, more intense, and burning areas previously thought untouchable. The gradual, relentless creep of sea-level rise began to force uncomfortable decisions in places like Miami, Venice, and Jakarta, but responses were often piecemeal–higher seawalls here, abandoned neighborhoods there–lacking the systemic overhaul required. Each disaster was met with expressions of shock and sympathy, followed by rebuilding efforts that largely replicated the vulnerabilities of the past. The "new normal" was accepted with a collective shrug, the baseline shifting imperceptibly year by year.

Agriculture felt the strain early on. Subtle shifts in rainfall patterns, hotter summers, and the spread of pests into new territories began impacting yields in major breadbaskets. The Ogallala Aquifer beneath the American Great Plains, relentlessly pumped for decades, showed alarming depletion rates, threatening a cornerstone of US agriculture. Similar stories unfolded around the world's vital groundwater reserves. Food prices began a volatile climb, hitting vulnerable populations hardest and occasionally sparking localized unrest, foreshadowing the widespread famine to come.

Amidst this growing crisis, a persistent narrative of techno-optimism offered a convenient excuse for delaying difficult policy choices. Grandiose promises of future technologies–fusion power just around the corner, city-sized carbon capture facilities, fleets of atmospheric geoengineering aircraft–captured the imagination and provided cover for inaction. While research continued, the breakthroughs needed to deploy these solutions at the necessary scale and affordable cost remained elusive. This faith in a technological *deus ex machina* allowed societies to avoid confronting the uncomfortable truth: the primary solution, stopping fossil fuel consumption and leaving it in the ground, was available but politically unpalatable.

Societal attitudes towards the crisis remained deeply fractured. While awareness of global warming was widespread, it often translated into a confusing mix of anxiety, fatalism, and cognitive dissonance. The sheer scale of the problem felt overwhelming, leading many to tune out or compartmentalize. Media coverage, often driven by short news cycles and a tendency towards "balanced" reporting that gave undue weight to fringe skepticism struggled to convey the urgency and complexity of the situation effectively.

The fragmentation of media into echo chambers further entrenched divisions. Individuals could curate news feeds that reinforced their existing biases, making constructive dialogue across ideological lines nearly impossible. Online platforms became battlegrounds for information warfare, where climate disinformation thrived alongside conspiracy theories about globalist plots or stolen elections. This erosion of shared reality was a critical precursor to the collapse of trust in institutions and experts, paving the way for neofascist leaders who offered overly simplified, incomplete answers to complex problems. There was also a pervasive sense of "disaster fatigue". As extreme weather events became more frequent, they risked losing their power to shock. Each new record-breaking storm or heatwave was assimilated into the background noise of a world seemingly beset by perpetual crises. This normalization of the abnormal bred a dangerous complacency, a sense that humanity could somehow muddle through, adapting incrementally to ever-worsening conditions without fundamentally changing course.

The decline wasn't a sudden plunge but a long, grinding descent. It was the slow erosion of Arctic Sea ice, year after year. It was the gradual acidification of the oceans, largely invisible but steadily undermining marine ecosystems. It was the incremental increase in global average temperatures, fractional degrees adding up with relentless consequence. It was the steady accumulation of political compromises that favored short-term economic gain over long-term survival. It was the countless missed opportunities to pivot towards a sustainable path.

Looking back from the ravaged landscape of 2050, the trajectory seemed almost predetermined. The political failures, the economic shortsightedness, the societal divisions, the environmental degradation–they were all interconnected, reinforcing each other in a positive feedback loop. The rise of authoritarian figures, the rejection of international cooperation, the dismissal of scientific warnings–these were symptoms of a deeper malaise, a failure to reckon with the biophysical limits to growth of human society on the planet and the political challenges of collective action.

The shadows of the past weren't just memories; they were active forces shaping the present. The infrastructure built around fossil fuels, the political systems resistant to change, the ingrained habits of over-consumption, the historical grievances between nations–all these legacies constrained the choices available and made the path towards collapse far easier to tread than the path towards recovery. The world didn't stumble blindly into the chaos of the mid-twenty-first century; it walked towards it with its eyes wide open, haunted and ultimately crippled by the ghosts of its own history. The decisions made, and not made, in the decades before 2050 cast long, dark shadows that stretched all the way to the grim realities of 2050.

CHAPTER 3

The ghosts of treaties past haunted the fractured world of 2050. They weren't spectral apparitions, but rather the tangible, grinding consequences of promises broken and frameworks abandoned decades earlier. The Paris Agreement and the North Atlantic Treaty Organization weren't just historical footnotes discussed by scavengers huddled around flickering fires; they were the gaping holes through which stability, cooperation, and hope had irrevocably drained. Their collapse in 2040 hadn't merely triggered the initial descent into chaos; the ongoing absence of anything remotely resembling them actively perpetuated the nightmare. They were the forsaken architecture of a world that might have been, their ruins serving as treacherous foundations for the precarious present.

The Paris Agreement, even in its prime, had been criticized as insufficient, a collection of voluntary pledges lacking teeth. Its collapse signaled a global surrender. The already weak dams holding back the floodwaters of climate catastrophe weren't just breached; they were systematically demolished. Nations, freed from even the pretense of collective responsibility, engaged in a desperate, short-sighted scramble for remaining resources, prioritizing immediate economic survival over any long-term planetary considerations. The fossil fuel industry, sensing its final, highly profitable twilight, ramped up extraction wherever politically feasible, often in ecologically devastating ways, locking in decades more of GHG emissions.

The positive feedback loops, once discussed in hushed tones at scientific conferences, began to operate with terrifying quickness. The rate of Arctic permafrost thaw accelerated dramatically through the 2040s releasing large amounts of methane (CH_4), a greenhouse gas far more potent than CO_2, that had been locked away for millennia. Methane is roughly 80 times more potent than carbon dioxide over a 20-year period and about 28 times more potent over a 100-year period. This wasn't a gentle leak; it was a geologically significant outgassing event, pushing global temperatures higher faster than even the more pessimistic pre-2040 models had predicted. The melting Greenland and Antarctic ice sheets contributed not just to sea-level rise, but also disrupted ocean currents like the Atlantic Meridional Overturning Circulation (AMOC), leading to chaotic and unpredictable weather patterns, particularly in Europe and North America. The stabilizing influence of these massive ice bodies was lost much quicker than anticipated once the global commitment to temperature limits evaporated.

Regions that had invested heavily in climate adaptation based on pre-2040 projections found their efforts tragically inadequate. Seawalls built to withstand mid-century sea level rise projections were overtopped by actual sea level rise. Vast desalination plants, intended to combat drought in places like California and the Mediterranean, became useless monuments as the energy grids needed to power them collapsed under the strain of extreme weather, resource scarcity, and a retreat from renewable energy. Genetically modified crops, engineered for slightly warmer or drier conditions, withered under unprecedented heatwaves, or were washed away by floods of biblical proportions. The adaptation debt, the cost of adjusting to changes already locked in, ballooned beyond the capacity of even the wealthiest remaining enclaves. The abandonment of the Paris Agreement meant abandoning the shared effort to *limit* the required adaptation to something potentially manageable.

The psychological scar left by the forsaken Paris Agreement ran deep. There was a pervasive, corrosive cynicism about collective human endeavors. People remembered–or were told stories about–a time when nations at least *talked* about solving the climate crisis together. The formal rejection of that goal fostered a sense of profound betrayal, particularly among younger generations who inherited the mess. It fueled fatalistic philosophies, nihilistic cults celebrating the planet's fiery end, and a bitter resentment towards the past perceived as suicidally complacent. Attempts to forge new, regional climate pacts in the years after 2040 invariably failed, shipwrecked on the rocks of mutual suspicion and the sheer scale of the problem, which clearly required a global response that was no longer conceivable. Why trust a neighbor's promise when the world's superpowers had so spectacularly broken theirs?

Simultaneously, the disintegration of NATO carved a security vacuum across vast and critical regions. It wasn't just the loss of Article 5, which stated that an armed attack against one or more of the member states shall be considered an attack against all, collective defense guarantee, which had become largely symbolic by the early 2030s anyway. It was the disappearance of the integrated command structures, the shared intelligence networks, the standardized equipment and procedures, and perhaps most importantly, the constant dialogue and joint exercises that had kept old European rivalries mostly dormant for nearly a century. When the American linchpin was definitively removed, the structure didn't just weaken; it shattered.

Europe, facing the brunt of climate migration from Africa and the Middle East, as well as its own internal climate stresses (drought in the south, chaotic weather in the north), fractured along old and new fault lines. Without the NATO framework to mediate disputes or coordinate responses, tensions over resources–particularly water and arable land–escalated rapidly. Border skirmishes became common. Nations invested dwindling resources in purely national defense, often directed against former allies. The dream of a unified, peaceful Europe evaporated, replaced by a patchwork of suspicious, heavily armed states and ungoverned territories.

The North Atlantic and the newly accessible Arctic became arenas of unchecked competition. Oligarchic states, sensing opportunity in NATO's collapse, sought to assert dominance over Arctic shipping routes and resources laid bare by melting ice. Other Arctic nations, lacking the collective security umbrella, responded with their own militarization efforts, leading to tense standoffs and occasional clashes over fishing rights, mineral claims, and navigational passage. Piracy, once largely suppressed by coordinated naval patrols, saw a resurgence, particularly along vital coastal routes where national coast guards were overwhelmed or non-existent. Shipping became significantly more dangerous and expensive, further crippling global trade.

The physical remnants of NATO became dangerous liabilities. Vast arsenals of weapons and ammunition, stored in bases across Europe, were often poorly secured as central command structures dissolved. Some were looted by retreating national armies, others fell into the hands of local militias, warlords, or organized crime syndicates. Sophisticated weaponry, from anti- aircraft missiles to armored vehicles, proliferated into conflict zones both within Europe and beyond, dramatically escalating the lethality of regional wars. The very tools designed to ensure peace through deterrence became instruments of chaotic violence. Bases themselves, like the missile facility on Crete, were either abandoned, repurposed by local factions, or fell into strategic disrepair, their sophisticated systems decaying without maintenance or oversight.

The absence of NATO's logistical and coordination capabilities hampered responses to the escalating climate disasters. In the past, NATO assets–heavy lift aircraft, engineering battalions, medical teams, satellite reconnaissance–had often been deployed to assist member states during floods, earthquakes, or major storms. This capacity vanished. Each nation, each region, was left to face catastrophic events alone, further eroding public trust in any form of large-scale governance and reinforcing the sense of abandonment.

The true catastrophe lay in the *synergistic failure* of both the Paris Agreement and NATO. The environmental chaos unleashed by the abandonment of the Paris Agreement directly fueled the conflicts that the defunct NATO could no longer prevent or contain.

Climate-driven resource scarcity—water shortages, crop failures, collapsing fisheries— became potent drivers of war. Mass migrations, triggered by lands becoming uninhabitable, overwhelmed borders and stoked xenophobia, providing fertile ground for nationalist demagogues and violent conflict. In turn, the pervasive insecurity and warfare made any large-scale climate mitigation or adaptation efforts utterly impossible. Who could invest in massive solar farms or coastal defenses when the territory might be overrun tomorrow? How could nations cooperate on managing shared river basins when they viewed each other as existential threats?

Farmers struggled to irrigate fields with dwindling river flow, while simultaneously trying to defend their communities from raids. There was no central authority to manage water resources equitably, no security force to maintain order, and no prospect of implementing the kind of large-scale water infrastructure projects that might have offered a long-term solution. The failure of the Paris Agreement exacerbated the drought; the failure of NATO allowed the ensuing chaos to fester unchecked.

This interconnected breakdown fostered a deep-seated, almost instinctual mistrust that permeated society by 2050. Treaties, alliances, international law— these concepts were regarded with derision, relics of a naive era when people pretended promises between nations meant something. Diplomacy, where it occurred, was ruthlessly transactional, based on immediate leverage and the credible threat of force. The idea of shared sacrifice for a common good, the very principle underpinning both the Paris Agreement and NATO, seemed totally alien. The lesson learned from 2040 wasn't that cooperation was difficult; it was that cooperation was for fools, and betrayal was the only rational strategy in a world without rules.

A former German Army officer who served in a joint NATO command at a surface-to-air missile site in northern Bavaria recalled the intricate web of communication, the trust built over years of joint exercises, the shared understanding of procedures. After 2040, he watched it all unravel. Units were recalled, intelligence feeds cut off, long-standing relationships severed by political decree. He saw former comrades-in-arms suddenly viewing each other with suspicion across newly hardened borders. Later, trying to organize local defense in his own town against resource raiders, he lamented the loss of that coordination, the inability to call for support or share reliable information beyond his immediate vicinity. The sophisticated tools of collective security had rusted away, replaced by patchwork militias armed with scavenged gear and fueled by fear.

The forsaken agreements represented more than just failed policies; they symbolized a deliberate choice by humanity, led by its most powerful actors, to turn away from the path of cooperation precisely when it was most needed. They were monuments to a colossal failure of foresight and political will. The consequences weren't just measured in degrees Celsius, meters of sea-level rise, or casualty counts from petty wars. They were measured in the pervasive despair, the endemic mistrust, and the near-total absence of any credible framework for rebuilding a functional global society. By 2050, the ghosts of the Paris Agreement and NATO served as constant, grim reminders that the world hadn't just stumbled into chaos; it had actively dismantled the very structures that might have offered a way out. The agreements were forsaken, and in turn, much of humanity felt forsaken itself, left to navigate the ruins alone.

CHAPTER 4

The Akrotiri Peninsula jabs a rocky finger into the Sea of Crete, a landscape sculpted by sun, wind, and millennia of stubborn history. Near its south eastern edge, nestled among thorny scrub and overlooking the turquoise expanse, sat the decaying carcass of the NATO Missile Firing Installation, known locally, back when such things mattered, simply as NAMFI. Deactivated officially sometime in the late 2030s as the Alliance crumbled, it hadn't truly been abandoned. Instead, it had burrowed deeper into secrecy, becoming the unlikely crucible for humanity's most audacious, perhaps final, gamble.

Outwardly, the base projected an image of utter dereliction. Rusting radar domes gaped blindly at the sky, missile launch gantries stood like skeletal remains picked clean by salty air, and perimeter fences sagged, breached in places by wandering goats or desperate scavengers who rarely found anything worth the trouble. The occasional drone overflight by one regional faction or another registered only silence and decay. This carefully cultivated neglect was the first layer of security for the project housed deep within Skloka Mountain's embrace.

Below the sunbaked surface, down shafts originally designed to withstand conventional ordnance and facilitate rapid missile deployment, life hummed with a quiet intensity. This subterranean warren, repurposed and reinforced, housed the Chania Facility–not its official name, as it had none, but the informal designation used by the handful of souls who knew of its existence. Here, shielded from the chaos above by meters of rock and layers of electronic countermeasures, Dr. Katy Belgrade and her small team waged their clandestine war against the fundamental laws of the universe.

Belgrade, a woman of 50 whose weary eyes held the weight of collapsing stars, was the intellectual heart of the operation. Once a celebrated theoretical physicist in Geneva, Switzerland before the funding dried up and international collaboration became a quaint memory, she possessed a mind capable of navigating the treacherous landscape where quantum mechanics melded into thermodynamics. She earned her PhD in quantum physics at the Massachusetts Institute of Technology (MIT) and was a native American tribal member from North Dakota, USA. She wasn't driven by ambition anymore. Accolades of the old world meant nothing. She was driven by a cold, terrifying clarity: the second law of thermodynamics, the universe's unwavering march towards disorder and heat death, was choking the Earth, amplified by humanity's technological addiction. And she, against all odds, thought she saw a loophole.

Her second-in-command, Sophia Solo, was the pragmatist who kept the lights on, metaphorically and literally. A former eastern European aerospace engineer with a genius for improvisation, she coaxed life out of aging generators, maintained the vital atmospheric seals, jury-rigged cooling systems for their power-hungry experimental apparatus, and managed the facility's meager, often perilously obtained, resources. Where Belgrade saw elegant equations, Solo saw potential points of failure, resource bottlenecks, and the sheer, grinding physics of keeping their fragile sanctuary operational in a world falling apart. Her pessimism was legendary, her competence indispensable.

Rounding out the core physics team were Cerise Montane, a quiet, meticulous experimentalist whose hands possessed an almost preternatural steadiness, essential for manipulating the delicate quantum states at the heart of their work, and James Ekota, a young, brilliant theorist whose insights often provided the conceptual leaps needed to bridge Belgrade' grand vision with Montane's tangible experiments. Ekota, barely thirty, represented the generation that had never known a stable world; his motivation was less about restoring the past and more about forging any kind of future at all.

Supporting them was a skeleton crew of technicians and security personnel, fewer than twenty souls in total. Most were former military or intelligence operatives, handpicked for their loyalty, discretion, and ability to survive in hostile environments. They maintained the physical security, ran silent patrols in the surrounding hills, managed the encrypted, sporadic communications bursts with their unseen patrons, and handled the logistics of acquiring supplies–a dangerous task involving clandestine meetings, black markets, and navigating the treacherous politics of the Aegean factions. Their leader, a grizzled ex-NATO operative known only as Nathan, enforced discipline with quiet authority, understanding that a single security breach meant not just project failure, but likely death for them all.

The Chania Directive wasn't a specific written order, but rather the implicit, all-consuming mandate under which they operated: achieve localized entropy reversal on a scale sufficient to impact Earth's climate system. Prove the theory, build a prototype, and demonstrate feasibility. It was a goal so audacious it bordered on impossible, yet they pursued it with the methodical rigor of engineers troubleshooting a faulty circuit. There was little room for error or doubt because the accelerating climate catastrophe provided a relentless deadline. However, as the physicist Richard P. Feynman said, "we absolutely must leave room for doubt or there is no progress and there is no learning." So, the team was sandwiched between a tremendous urgency and some doubt as to how to move forward because they could not know everything they needed to know. There would be trial and error…and luck involved.

Secrecy was woven into the very fabric of their existence. The facility's true power signature was masked, buried beneath layers of electronic noise designed to mimic geothermal activity or defunct military equipment.

Supplies arrived irregularly, often disguised as salvage hauls, or delivered via submersible drones to hidden coves along the coast. Personnel rotation was non-existent; everyone inside was committed for the duration, however long that might be. Contact with the outside world was filtered through Nathan and limited to heavily encrypted, need-to-know transmissions with the shadowy remnants of whatever organization still funded and protected them–likely a splintered faction from a defunct intelligence agency or a forward-thinking element of a collapsed government clinging to a final, desperate hope.

Life underground was a monotonous cycle punctuated by moments of intense intellectual fervor or system-failure panic. The recycled air carried the tang of ozone and machine oil. Artificial lighting mimicked diurnal rhythms, but the absence of true sun and open sky wore on the inhabitants. Food was nutritious but repetitive, mostly hydroponically grown staples supplemented by carefully rationed preserved goods. Water was meticulously recycled. Personal space was minimal, privacy a forgotten luxury. The constant low hum of machinery was the soundtrack to their lives.

Their laboratory was a bizarre fusion of salvaged Cold War infrastructure and innovative, custom-built experimental physics equipment. Thick blast doors separated sections designed to contain reactor cores or missile engines, now repurposed to house delicate quantum entanglement arrays and resonant field generators. Cables snaked like metallic vines across floors and ceilings, connecting legacy power conduits to hyper-specialized sensors. Belgrade often joked that they were trying to rewrite the laws of physics using spare parts from a museum of mutually assured destruction.

Progress was agonizingly slow, measured in fractions of degrees Kelvin difference achieved in microscopic volumes for fleeting nanoseconds. Belgrade would spend days locked in simulation, manipulating variables in the complex models that described the negentropic effect they sought to harness. Ekota would pore over Belgrade's outputs, searching for flaws or alternative interpretations, occasionally emerging with a breakthrough insight that sent Montane scurrying back to the apparatus to recalibrate or redesign a crucial component. Solo, meanwhile, battled entropy in its more mundane forms: failing capacitors, coolant leaks, power fluctuations from their aging primary reactor – a small, decommissioned naval unit they kept running through sheer ingenuity and cannibalized parts.

The psychological pressure was immense. They were working in complete isolation, cut off from news of loved ones, knowing the world above was burning. The weight of their potential success was as heavy as the fear of failure. What if they succeeded? How could such a technology be controlled? Deployed? Could it truly save the world, or would it merely become another weapon in the hands of the powerful? These questions sometimes surfaced in hushed conversations during shared meals or late- night maintenance checks, but were usually pushed aside. The immediate task–making the damn thing work– consumed all available mental bandwidth.

Sometimes the outside world intruded. A tremor from a distant conflict, felt deep within the rock. A garbled radio transmission hinting at escalating famine in North Africa. A failed supply rendezvous, forcing Solo to implement even stricter rationing. Once, a scavenger crew, more persistent or luckier than most, stumbled upon a hidden ventilation shaft. Nathan's security team dealt with the intrusion silently and efficiently, a grim reminder of the stakes and the ruthlessness required to protect their secret. The incident cast a pall over the facility for weeks.

Despite the hardships and the glacial pace, there were moments of fragile hope. A successful test run, however small the scale, where Montane's instruments registered a sustained, albeit tiny, decrease in entropy within the containment field. A new simulation from Belgrade predicting a more stable pathway to amplification. A salvaged piece of advanced computational hardware, acquired at great risk by Nathan's team, that significantly sped up their modeling capabilities. These small victories were celebrated quietly, with rationed coffee and shared relief, before the relentless work resumed.

They rarely interacted directly with the local Cretan population. Chania, though relatively close geographically, existed in a different universe. The city, like many Mediterranean ports, struggled under the weight of climate refugees, resource scarcity, and the weak, often corrupt, governance of local factions. Occasionally, Nathan's operatives bartered for fresh produce or specific medical supplies in the Chania markets, operating under deep cover, but direct contact was strictly forbidden for the core team. They were ghosts, living and working beneath a land already haunted by layers of history, now adding their own secret chapter.

The Chania Directive demanded results, but the universe was stingy with its secrets. Belgrade knew they were probing forces barely understood, flirting with energies that could potentially unravel spacetime itself if mishandled. The line between reversing entropy and creating some new, unforeseen catastrophe was terrifyingly thin. Were they saviors or sorcerer's apprentices? She pushed the thought away, focusing on the next equation, the next simulation run. Solo worried less about cosmic consequences and more about the primary heat exchanger failing during a critical experiment. Montane meticulously documented every minute energy fluctuation, her focus absolute. Ekota dreamed of stable negentropic fields, occasionally waking with a solution that seemed brilliant until subjected to the harsh light of calculation.

The Chania Facility was an island of high technology and desperate intellect adrift in an ocean of global decay. It was a microcosm of humanity's contradictory nature: the capacity for profound insight coupled with the legacy of self-destructive tendencies, the drive to create miracles housed within the architecture of war. Whether the Directive would lead to salvation or simply provide a more spectacular final act for a species consumed by its own runaway entropy remained, like the future of the planet itself, agonizingly uncertain. But deep beneath the Cretan soil, the work continued, shielded by rock and secrecy, against the dying light of the world.

CHAPTER 5

The fundamental problem wasn't merely the heat, the rising waters, or the ferocious storms that ravaged the Earth of 2050. Those were symptoms. The disease, as Dr. Katy Belgrade understood it with chilling clarity deep beneath the Cretan mountains, was *entropy*. The relentless, universal tendency of systems to progress from order to disorder, for energy to dissipate, for heat to spread until everything reaches a state of uniform, tepid equilibrium. Global warming was just entropy writ large on a planetary scale, accelerated by humanity's prodigious talent for unlocking ancient carbon stores and releasing their heat trapping effects which impacted the atmosphere and oceans. To save the Earth, Belgrade believed, one had to attack the disease, not just the symptoms. One had to find a way to make entropy run backwards, even if only locally—to create *negentropy* or low entropy.

The second law of thermodynamics, the bedrock principle codifying this inexorable march towards disorder, was perhaps the most inviolable tenet in classical physics. It dictated the arrow of time, explained why dropped eggs don't spontaneously reassemble, and why perpetual motion machines remain firmly in the realm of fantasy. Heat naturally flows from hot to cold, never the other way around without inputs of external energy. Complex structures decay; concentrated energy disperses. Reversing this, forcing heat to flow from the cooler atmosphere back into a concentrated, manageable state, or simply making disorganized energy become organized again, seemed not just technologically impossible, but fundamentally contrary to the nature of reality. It was akin to demanding that a scrambled egg unscramble itself back into its shell.

Yet, Belgrade wasn't dealing purely with classical physics. Her domain was the murky, counter-intuitive world of quantum mechanics, where probabilities replaced certainties and effects could seemingly precede causes. Decades earlier, before the world truly began its nosedive, theoretical discussions occasionally surfaced about loopholes–Maxwell's famous Demon sorting molecules, quantum entanglement potentially allowing information or energy transfer in non-classical ways, Schrodinger's cat, and the bizarre nature of vacuum energy. Most were thought experiments, mathematical curiosities. But Belgrade, driven by the looming catastrophe and gifted with an intellect that saw connections others missed, believed she had found more than a curiosity. She had found a potential mechanism, a way to, as she dryly put it, "politely request the universe to tidy up after itself, just in this one small corner."

Her hypothesis revolved around the concept of precisely tuned resonant energy fields interacting with quantum vacuum fluctuations. The vacuum, far from being empty, teemed with virtual particles popping in and out of existence, a sea of fleeting potential. Belgrade theorized that by generating an extremely complex, multi-layered resonant field – analogous to striking a vast, multi-dimensional tuning fork – they could selectively influence these vacuum states within a contained area. The goal wasn't simply to extract energy from the vacuum, a common trope of older science fiction, but something far more subtle: to use the field to preferentially dampen certain energetic states associated with disorder (high entropy) while amplifying states associated with order (low entropy). In essence, the field would act as a filter or guide, nudging the local quantum field towards configurations that represented a lower overall entropy state for the matter and energy within the field's influence.

The practical application, if it could be achieved, was staggering. Such a field could theoretically coax disorganized heat energy in the atmosphere within its bounds to coalesce, to become more concentrated, effectively cooling the targeted volume. It wouldn't destroy the energy, violating the first law of thermodynamics (conservation of energy), but it would rearrange it, reducing the local disorder, seemingly defying the second law. Where the excess, now more ordered, energy would go was one of the terrifyingly large theoretical questions. Would it be shunted into hyperspace? Radiated away in a controlled burst? Or perhaps converted into exotic particles? Belgrade's models suggested several possibilities, none entirely comforting.

Translating this elegant, if terrifying, theory into practice fell largely upon Cerise Montane's shoulders. The heart of the Chania Facility's experimental efforts was the Resonant Field Core; a monstrosity of engineering housed in a chamber originally designed to withstand a direct missile attack. It was a spherical lattice of superconducting magnets, exotic metamaterial emitters, and hyper-sensitive detectors, all converging on a central containment zone barely a meter across. This zone was where the magic–or the catastrophe–was supposed to happen. The complexity lay in generating the precise, dynamic field geometry Belgrade's theories demanded. It required coordinating thousands of individual emitters, adjusting frequencies, amplitudes, and phase relationships on timescales measured in picoseconds (10^{-12}), all while bathed in the intense energies needed to interact with the quantum vacuum.

Sophia Solo's constant battle with the facility's aging power systems was directly linked to the Core's voracious appetite. Generating the resonant fields required bursts of power equivalent to a small city's consumption, placing immense strain on their salvaged naval reactor and energy storage systems. Equally challenging was the cooling; manipulating energy fields on this level generated colossal amounts of waste heat–a cruel irony, given their goal. Solo's jury-rigged heat exchangers and cryogenic systems groaned under the load during every test run, constant reminders of the practical friction opposing Belgrade's frictionless equations.

Early experiments, conducted over the preceding few years, had been both tantalizing and frustrating. They started small, focusing on creating negentropic effects within microscopic volumes of inert gas. Montane, with painstaking precision, would align the emitters, initiate the power sequence, and watch the sensor readings. Occasionally, for fleeting moments – nanoseconds (10^{-9}), sometimes femtoseconds (10^{-15})–the instruments would register the impossible: a localized temperature drop violating classical heat flow, a brief, statistically significant decrease in the measured entropy of the gas molecules within the containment field. These were victories measured in decimal points, achieved at enormous energy cost, but they were victories nonetheless. They proved the principle wasn't entirely theoretical madness.

The primary obstacle was coherence. Belgrade's resonant fields required near-perfect synchronization and stability. The slightest external vibration, power fluctuation, or instability within the plasma containment could disrupt the delicate resonance, causing the negentropic (low entropy) effect to collapse instantly. Montane compared it to trying to build a sandcastle during an earthquake while simultaneously balancing it on the head of a pin. Maintaining the field coherence long enough and stable enough to achieve a measurable, sustained effect, even within their small test volume, proved maddeningly difficult. Each successful nanosecond felt like a hard-won battle against the universe's inherent chaos.

James Ekota, the youngest of the core team, often played devil's advocate, probing Belgrade's theories for weaknesses and exploring the potential downsides predicted by the mathematics. While Belgrade focused on achieving the primary effect, Ekota worried about the second-order consequences. Did the equations predict unforeseen quantum tunneling events? Could the resonant field inadvertently create micro-singularities or exotic particles that couldn't be contained? His simulations sometimes hinted at strange resonances, feedback loops that could potentially amplify uncontrollably under certain conditions. "We're tickling a sleeping dragon," Ekota once remarked grimly after a simulation run produced anomalous energy spikes. "We hope it purrs, but it might wake up hungry."

One particular experiment, designated Test Run Alpha Bravo Niner, aimed to push the boundaries slightly further. The goal was to sustain a measurable negentropic effect within a one-liter volume of standard atmosphere (carefully filtered and stabilized) for a full microsecond – an eternity compared to their previous nanosecond successes. The power requirements were immense, pushing Solo's generators to their absolute limit. The calculations for the field geometry were the most complex yet, taking weeks of debate between Belgrade and Ekota, and refined by Montane's practical insights into the Core's limitations.

The countdown sequence was tense, the underground lab hushed except for the hum of machinery and Solo's clipped status reports over the intercom. Montane monitored the emitter array diagnostics, her face impassive. Belgrade watched the primary entropy sensor feed, her knuckles white. Ekota tracked simulated field stability against real-time readings, ready to hit the emergency shutdown. The power surged, dimming lights outside the shielded control room. The resonant hum from the Core deepened, shifting pitch as the complex field patterns stabilized.

For a fraction of a second, everything seemed nominal. Then, the entropy reading dipped sharply–success! A cheer started to rise, instantly stifled. The dip continued, plunging faster than predicted. Simultaneously, strange Cherenkov radiation signatures flared on secondary sensors, and the containment field monitoring systems registered a brief, localized spatial distortion–a flicker in the fabric of space itself. Alarms blared. Montane initiated an emergency field collapse sequence. The resonant hum died with a sickening clang of protesting machinery.

Silence descended, broken only by the whine of emergency cooling systems. The test vessel was intact, the Core stable, but the data was deeply unsettling. They had achieved the microsecond duration, even exceeded the target entropy reduction momentarily. But the unexpected radiation signature and the spatial flicker were entirely outside their models. Ekota was pale. "The resonance... it almost coupled with something else. Something fundamental. The models don't account for this."

Belgrade stared at the data, her initial flicker of triumph extinguished as she said "The field geometry wasn't perfect. There was a harmonic... an instability we didn't predict at this energy level." She looked not defeated, but intensely focused. "We forced the local system towards order, but the energy cost, the stress on the local vacuum... it had to manifest somehow. This is new territory."

The incident highlighted the profound risks inherent in their work. They weren't just building a machine; they were probing the very structure of reality, and reality, it seemed, might push back in unpredictable ways. Could scaling up the effect trigger larger, uncontrollable spatial or temporal anomalies? Was there a fundamental limit to how much order could be imposed before the universe's underlying framework buckled? These questions hung heavy in the recycled air of the facility.

Furthermore, the energy cost remained a monstrous hurdle. Even their most successful micro-experiments required staggering amounts of input power to achieve minuscule reductions in entropy. The paradox was stark: they were using vast amounts of low entropy energy (from their reactor) to impose a tiny amount of order locally. While the ultimate goal was to harness the effect to cool the *entire planet*, requiring a net energy removal, the initial activation energy to establish and maintain the negentropic field was colossal. Solo constantly reminded Belgrade that scaling up would require power sources far beyond their current capabilities, potentially requiring technologies as revolutionary as the entropy reversal itself.

Where did the energy extracted from the cooled volume actually go? Belgrade's preferred model suggested it was shunted into higher-dimensional space, effectively removed from their spacetime framework. Ekota worried it might be accumulating somewhere, creating an entropy debt that could come due catastrophically. Montane, ever the pragmatist, focused on improving the efficiency of the field emitters, trying to reduce the sheer waste heat and power draw, regardless of the ultimate destination of the extracted energy.

Despite the dangers and the daunting challenges, Test Run Alpha Bravo Niner was considered a qualified success. It proved that a sustained, measurable reduction in entropy was possible beyond the nanosecond scale. It validated the core principles of Belgrade's resonant field theory, even as it exposed gaps in their understanding and highlighted new risks. The path forward involved meticulous analysis of the anomalous data, refining the field geometries to avoid dangerous harmonics, improving the stability and efficiency of the Core, and grappling with the terrifying implications of scaling up.

The science of reversal was not a matter of flipping a switch. It was a painstaking, perilous exploration into the deepest laws of physics, conducted under immense pressure and in profound isolation. Belgrade, Solo, Montane, and Ekota were not just engineers; they were undaunted explorers charting a course through utterly unknown territory, guided by equations that hinted at both success and oblivion. The fate of a world burning up in its own heat hung on their ability to navigate the treacherous interface between quantum mechanics and thermodynamics, to master the delicate art of coaxing order out of chaos, without accidentally tearing reality apart in the process. The whispers of their secret work might be faint, but the forces they wrestled with in the heart of the Chania Facility were anything but.

CHAPTER 6

The drone banked silently against the bruised twilight sky, its thermal camera scanning the desolate coastline below. Miles away, huddled in a makeshift command post carved into a hillside overlooking the Aegean, JoAnn Solo watched the feed on a flickering, salvaged tablet. Her calloused fingers traced the outline of a hidden cove, cross-referencing it against encrypted coordinates received hours earlier via a burst transmission that had bounced off three different defunct satellites. Tonight's delivery was critical: superconducting wire, desperately needed replacement components for a cryogenic cooling system, and, incongruously, three kilograms of high-quality coffee beans–a specific request from Nathan, relayed with unusual urgency.

JoAnn wasn't military, not officially. Nor was she strictly intelligence, though her skillset overlapped considerably. She was an operative for the entity known only by its internal designator: OPSEC. OPSEC wasn't a faction in the conventional sense. It held no territory, flew no flags, and its existence was known only to its own compartmentalized cells and, presumably, to whatever shadowy figures pulled the strings from even deeper obscurity. Born from the wreckage of several pre-collapse supra- national intelligence and strategic foresight agencies, OPSEC represented a desperate attempt by a handful of forward-thinking individuals to preserve not just data, but capability–the ability to identify existential threats and potentially nurture long-shot solutions in a world consumed by short-term survival.

The Chania Directive, the project helmed by Dr. Katy Belgrade deep beneath the NATO missile base, was OPSEC's longest shot, its most guarded secret, and to some degree its purpose. JoAnn's cell, based somewhere in the chaotic Eastern Mediterranean, was tasked with providing logistical support and peripheral security monitoring for the Chania Facility. They were the invisible shield, the silent logisticians for humanity's last, improbable gamble. Their work involved navigating a minefield of competing factions, treacherous black markets, and the constant threat of discovery.

OPSEC operated on principles of extreme decentralization and need-to- know. JoAnn knew her immediate superior only as "Control", a disembodied voice filtered through layers of encryption. She managed a small team of specialists: pilots who could fly salvaged drones through hostile airspace, technicians who could maintain aging comms gear, and ground operatives who could move like ghosts through faction-controlled ports and barren landscapes to make clandestine deliveries or gather intelligence. She suspected other OPSEC cells existed, perhaps focusing on recruitment, funding, counter-intelligence, or even other high-risk projects, but she had no direct knowledge of them. Compartmentalization was absolute; a captured operative could only reveal their own small part of the puzzle.

The motivation driving OPSEC, as far as JoAnn understood it, was a blend of cold pragmatism and profound duty. Its founders, witnessing the collapse accelerate in the late 2030s, recognized that conventional governance structures were doomed. They believed humanity's only hope lay in preserving pockets of advanced knowledge and capability, shielded from the political chaos, and identifying unconventional solutions that bypassed the broken international system. Global warming was identified early as the overriding existential threat. While geoengineering schemes were deemed too politically volatile and likely to be weaponized, Belgrade's fringe theories on entropy reversal, initially dismissed by mainstream science, caught the attention of OPSEC analysts searching for truly paradigm-shifting possibilities. It was a desperate bet, placed when all other options seemed to be failing.

Supporting the Chania Facility was a logistical nightmare. Resources were obtained through a complex web of channels. Some funding likely came from pre- collapse endowments, hidden caches of precious metals, or digital assets somehow preserved through the financial meltdowns. But much relied on OPSEC operatives actively acquiring what was needed in the fractured world of 2050. This involved infiltrating black markets, bartering with scavenged technology, manipulating faction supply chains, and occasionally conducting deniable raids disguised as piracy or inter-factional skirmishes. An OPSEC front company, ostensibly dealing in maritime salvage, might acquire a decommissioned naval reactor component needed by Solo, routing it through three different ports under false manifests before it reached JoAnn's team for final delivery. A cell embedded within a corporate agri-faction might divert shipments of specialized lubricants or chemical precursors, marking them as "lost in transit".

The coffee beans for Nathan were a simpler, yet telling, challenge. A luxury item, almost impossible to find genuine, high-quality beans. It required activating a sleeper contact within a South American faction still managing a few high-altitude plantations, arranging payment through untraceable barter goods, and using a long-established courier network to transport the small, precious package halfway across the world, switching hands a dozen times before finally reaching JoAnn's staging area. The effort involved seemed disproportionate, but

OPSEC understood the psychological value of such things for maintaining morale in the crushing isolation of the Chania Facility. Sometimes, hope arrived in the form of caffeine.

Intelligence gathering was equally crucial. JoAnn's team constantly monitored factional activity around Crete. Shipping movements, aerial patrols, radio chatter, human intelligence gathered by operatives posing as fishermen or traders–all fed into a threat assessment matrix. They tracked resource scarcity in nearby settlements, understanding that desperation could drive factions to probe the seemingly abandoned NAMFI base more aggressively. A rumor picked up in a Chania taverna about increased pirate activity might trigger heightened surveillance of coastal approaches. A spike in encrypted communications from a regional warlord's headquarters could signal preparations for an offensive that might spill over towards Akrotiri. This information was relayed to Nathan, usually via coded, indirect warnings, allowing him to adjust the facility's security posture or emission masking protocols.

The disguise aspect was fundamental to OPSEC's survival. Its operatives blended into the background noise of the chaotic world. They were unremarkable traders, cynical mercenaries captaining tramp freighters, disillusioned technicians working for corporate enclaves, weary aid workers navigating refugee flows. Kim, Reuben, and Ben appeared to all be mid-level logistics coordinators, efficient but unremarkable, known for their ability to source scarce parts through unofficial channels. In reality, they subtly manipulated shipping manifests and maintenance schedules, diverting critical supplies towards OPSEC pickup points or creating blind spots in coastal surveillance grids during sensitive operations like JoAnn's drone flights. Their lives were a constant balancing act; a single mistake, a misplaced document, or an overly curious superior could expose them, leading to brutal interrogation and the potential compromise of their entire network. They lived with the gnawing fear that their carefully constructed cover might be blown.

Operating within OPSEC wasn't just dangerous; it was psychologically taxing. Decades of secrecy, isolation from loved ones (most operatives were recruited precisely because they had few remaining ties), and the constant proximity to betrayal took their toll. Maintaining loyalty within such a fragmented, high-stress environment was a continuous challenge for Control and the unseen leadership. Operatives were vetted rigorously, often monitored subtly themselves, and bound by a shared belief in the mission's importance–or perhaps, simply by the lack of any viable alternative in the ruined world. Burnout was common. Disappearances occurred; whether they were due to enemy action, internal security measures, or simply individuals succumbing to despair was often impossible to determine.

Internal disagreements surely existed within OPSEC's hidden hierarchy, though JoAnn only sensed echoes of them. Was the Chania Facility consuming too many resources? Were Belgrade's theories truly viable, or just elegant fantasy? Should they prioritize preserving other forms of knowledge, like advanced biological sciences or computational infrastructure, over the entropy project? These debates likely occurred in encrypted forums or face-to-face meetings in secure locations JoAnn couldn't even imagine. The fact that support for Chania continued suggested that, for now, the proponents of Belgrade's gamble held sway, perhaps bolstered by the increasingly dire climate projections and the lack of any other plausible path towards planetary-scale recovery.

Nathan, the grizzled security chief within the Chania Facility, was the crucial interface between Belgrade's team and their hidden benefactors. He received the coded supply requests, the threat warnings, and presumably, broader strategic directives from OPSEC via communication channels known only to him. He, in turn, provided filtered progress reports back to Control – carefully worded summaries scrubbed of overly technical details or sensitive vulnerabilities. This arrangement protected both sides. The scientists remained focused on their work, shielded from the messy realities of resource acquisition and factional politics. OPSEC maintained effective security, ensuring that even the core Chania team didn't know the full identity, structure, or capabilities of the organization keeping them alive. To Belgrade and her colleagues, Nathan was simply the man who made the impossible logistics happen, the quiet guardian who dealt with the harsh realities of the outside world so they didn't have to. They likely suspected he had powerful backing, but the specifics remained deliberately vague.

JoAnn watched on her tablet as a small, semi-submersible drone surfaced briefly in the designated cove, its manipulator arm extending to retrieve the waterproof cache deposited earlier by one of her ground operatives. The exchange was clean. The submersible slipped back beneath the waves, presumably heading towards a hidden underwater access point connected to the Chania Facility. Another delivery completed. Another small victory in a silent, global war few knew was being fought.

She allowed herself a brief moment of satisfaction before erasing the drone's telemetry logs and shutting down the encrypted feed. The work was relentless, thankless, and carried the constant risk of catastrophic failure.

Yet, JoAnn and her counterparts across the scattered OPSEC network continued. They were the stagehands working in the shadows, ensuring the actors on the main stage–Belgrade, Solo, Montane, Ekota–had what they needed to perform their potentially world-altering play. They operated on faith–faith in Belgrade's science, faith in OPSEC's mission, or perhaps just faith in the necessity of trying *something*, anything, against the overwhelming tide of global warming, entropy, and despair.

These allies in disguise, hidden within the folds of a collapsing world, represented a paradox. They were remnants of the old order–the intelligence apparatus, the strategic thinking, the belief in large-scale, clandestine projects– yet they operated entirely outside its failed structures. They wielded sophisticated tools and tradecraft but relied on ancient methods of smuggling and barter. They aimed to save humanity but operated with a ruthlessness born of necessity, prepared to sacrifice individuals or manipulate events for the greater good as they perceived it. Their existence was a testament to the enduring human capacity for organization and long-term planning, even amidst chaos, but also a chilling reminder of the lengths some were willing to go to, in secret, when the fate of the world hung in the balance. They were the whispers made manifest, the hidden hands attempting to guide destiny, their success or failure utterly invisible to the billions they sought, perhaps quixotically, to save.

CHAPTER 7

The hum was omnipresent. Not the resonant thrum of the experimental Core during a test run, which vibrated through the very rock of the mountain, but the constant, low-level symphony of life support: air recyclers scrubbing CO_2, water pumps circulating coolant, the faint electronic whine of servers processing environmental data and maintaining internal systems. It was the sound of survival, meticulously engineered, deep beneath the sun-scorched skin of Crete. For the inhabitants of the Chania Facility, this subterranean world was their entire reality, a hermetically sealed bubble against the chaos consuming the planet above. They were the keepers of a secret too vast and too dangerous for the fractured world to comprehend, and the weight of that knowledge pressed down as surely as the meters of rock above their heads.

Dr. Katy Belgrade often sought refuge in the abstract beauty of her equations, the whiteboard in her small, spartan quarters covered in elegant symbols that described forces capable of rewriting existence. Yet, even here, the human element intruded. She traced a complex tensor notation, but her mind drifted to the face of Cerise Montane during the unsettling Alpha Bravo Niner test run–that flicker of something beyond controlled physics. Belgrade, outwardly the detached visionary, carried the burden not just of the science, but of the lives entrusted to her, the lives gambled on her theories. She hadn't seen her daughter, Jane, assuming she still lived somewhere in the fractured remains of continental Europe, in nearly a decade. The Chania Directive had demanded absolute severance. Was the potential salvation worth the certain sacrifice? The equations offered no answer.

Sophia Solo found solace in the tangible. Her world was one of pipes, conduits, pressure gauges, and power readings. While Belgrade wrestled with quantum dynamics, Sophia wrestled with failing gaskets and overheating capacitors. She moved through the facility's service tunnels with a practiced, economical gait, her toolkit a familiar weight on her hip. Her skepticism, often mistaken for cynicism, was a shield forged in the harsh realities of resource scarcity and equipment decay. Every successful day was a victory against entropy in its most mundane, relentless form. She knew the facility's weaknesses intimately–the aging reactor core borrowed from a decommissioned submarine, the temperamental atmospheric processors salvaged from a failed Mars colonization project, the finite stockpile of irreplaceable components OPSEC operatives risked their lives to deliver. She trusted Belgrade's intellect, but she trusted thermodynamics more. Keeping the lights on, the air breathable, and the

Core from melting down–that was her directive.

Cerise Montane maintained her equilibrium through meticulous routine and quiet discipline. Her small living space was immaculate; her few personal possessions arranged with geometric precision. In the lab, her hands moved with unwavering steadiness, calibrating sensors sensitive enough to detect the whisper of quantum fluctuations. She rarely spoke of the past, but fragments sometimes surfaced–memories of a traditional garden in Kyoto, the scent of cherry blossoms, a world of subtle order utterly unlike the humming confinement of the Chania Facility. She channeled his focus entirely into the experimental process, finding a kind of Zen wisdom in the pursuit of verifiable data. The anomalies, the unpredictable flickers like the one in Alpha Bravo Niner, disturbed her innate desire for repeatable results, hinting at forces less predictable than the equations suggested. Her silence wasn't emptiness, but a deep well of controlled concentration.

James Ekota, the youngest, provided a restless energy that contrasted with the weary determination of the others. He bounced between theoretical simulations and practical problem-solving, often challenging Belgrade's assumptions or proposing modifications to Montane's experimental setups. Having come of age during the collapse, he lacked the others' visceral memory of the world they were trying to save, but possessed a fierce idealism nonetheless. He saw the project not just as a scientific challenge, but as a moral imperative. Yet, his youth also made him more vulnerable to the psychological strain. He sometimes spent hours staring at the simulated sky projected onto the ceiling of the small communal area, a pale imitation of the real thing, the longing for open air and genuine sunlight etched on his face. He represented the future they fought for, and the generation most acutely aware of the stakes.

Overseeing the delicate human ecosystem, alongside Sophia's management of the physical one, was Nathan. The security chief moved through the facility with a quiet watchfulness, his eyes missing nothing. His past was a closely guarded secret, though his bearing and expertise hinted at years in elite special forces or deep-cover intelligence work. He was the gatekeeper, the filter through which all information from OPSEC flowed in, and carefully curated progress reports flowed out. He managed the small team of technicians and security personnel–the support crew–who performed the essential tasks of maintenance, perimeter monitoring (both physical and electronic), and dealing with the grim necessities of protecting their secrecy. Nathan understood the mission's importance, but his primary loyalty seemed focused on the survival and operational integrity of the facility and its personnel. He trusted Belgrade's science implicitly, not because he totally understood it, but because OPSEC deemed it critical. His job was to ensure the scientists had the secure environment they needed to work their marvel, or die trying.

The support crew formed a distinct social stratum within the bunker. Fewer than twenty individuals, mostly ex-military or technical specialists recruited by OPSEC for their skills and perceived loyalty, they lived slightly apart from the core science team. Their work was often physically demanding–patrolling the outer access tunnels, maintaining heavy machinery, managing waste recycling systems, running security drills. While the scientists debated quantum states, the support crew dealt with faulty wiring, strange noises in the ventilation shafts, or the logistics of receiving clandestine supply drops via the hidden sub-access point. Their motivations were often more pragmatic: a secure billet in a collapsing world, loyalty to Nathan or OPSEC, or simply a belief in performing their assigned duty. Communication with the scientists was typically professional and task-oriented, though mutual respect existed, born of shared confinement and dependency.

Life settled into a rhythm dictated by work schedules, energy availability, and the monotonous cycle of recycled air and artificial light. Meals were taken communally in a small mess hall, the food nutritious but repetitive–hydroponic vegetables, processed algae paste, occasional protein supplements. Conversation often revolved around work–troubleshooting technical issues, debating simulation results, planning the next experimental run. Personal discussions were more guarded, constrained by the need for operational security and the sheer weight of their shared isolation. Small rituals emerged: Sophia's fiercely competitive chess games with Montane (who usually won through sheer patience), Ekota's attempts to cultivate a small patch of herbs under a grow-lamp in his quarters, Belgrade's solitary walks through the quieter levels late at night. Nathan insisted on regular physical training for everyone, scientists included, utilizing a small, repurposed storage bay equipped with basic exercise gear–less for fitness, more for stress management and maintaining discipline.

Maintaining psychological health was a constant battle. The absence of natural light, the confinement, the awareness of the dying world above, and the immense pressure of their task created a potent cocktail of potential despair. OPSEC provided limited psychological support resources–encrypted access to therapy programs, curated entertainment files–but ultimately, resilience was an individual responsibility. Some relied on intense focus on their work, others on rigid routine, a few on carefully rationed vices like Nathan's requested coffee or hoarded pre-collapse chocolate. The ever-present knowledge that failure meant not just project termination but likely the end of any hope for humanity was a burden carried silently by all. They were keepers of the flame, but the bunker often felt more like a tomb.

Tensions inevitably arose. Belgrade's occasional theoretical leaps sometimes clashed with Solo's cautious engineering pragmatism. Ekota's impatience occasionally grated on Montane's methodical pace. The support crew sometimes chafed under restrictions imposed for scientific reasons they didn't fully grasp. Nathan mediated these frictions with quiet authority, reminding everyone of the larger goal and the need for absolute cohesion. Mistakes weren't just setbacks; they could be fatal, compromising the entire operation. The shared danger, paradoxically, often served to reinforce bonds, forging a unique camaraderie born of extreme circumstances. They were utterly reliant on each other, their individual skills complementing the whole.

The outside world remained a distant abstraction, filtered through Nathan's concise, sanitized intelligence updates provided by OPSEC. They knew of the major factions, the spreading wastelands, the general trajectory of collapse, but lacked the visceral, day-to-day understanding of life beyond their shielded walls. This isolation was necessary for security but also fostered a sense of disconnection. Were their efforts truly relevant to the struggles unfolding above? Could their potential solution even be implemented in such a fractured, distrustful world? These questions sometimes haunted the quiet hours, whispers of doubt in the hum of the machines.

They had all sacrificed immensely to be here. Families left behind, careers abandoned, identities erased. They were ghosts in the machine, their existence dedicated to a single, monumental task. They rarely spoke of these personal costs, burying them beneath layers of professional focus and disciplined routine. Yet, the ghosts remained, flickering in unguarded moments–a fleeting expression during a communal meal, a wistful glance at a faded photograph pinned above a workstation, a sudden silence in a conversation. They were the price of admission to this desperate, last-ditch effort, the personal sacrifices made in the hope of reversing the planet's entropy in an effort to combat global warming.

The secret they kept was twofold: the scientific secret of entropy reversal, and the human secret of their own fragile existence within the bunker. Protecting both required constant vigilance, unwavering discipline, and a degree of trust in each other that bordered on absolute, despite the inherent paranoia of their situation. They were prisoners of their own potential salvation, bound together by the immense gravity of their mission. The hum of the life support systems was the constant reminder of their isolation, their dependence, and the thin line they walked between breakthrough and oblivion, deep beneath the indifferent Cretan landscape. Each passing day was another small victory against failure, another day closer to an uncertain outcome, guarded by the few who dared to probe the universe's secrets.

CHAPTER 8

The sun over the Sea of Crete was a malevolent eye, burning through the thin, dusty haze that passed for atmosphere. The heat was a physical presence, pressing down, stealing breath, warping the horizon into shimmering mirages.

Chania emerged from the heat haze like a carcass picked clean by vultures and time. The old Venetian harbor walls were partially submerged, crumbling breakwaters barely disturbing the unnaturally calm, oily surface of the water. Buildings stood as hollowed-out shells, salt- encrusted and sun-bleached. The air hung heavy, thick with the stench of brine, decay, and something else... a low-level chemical taint lingering from industrial spills during the collapse.

Deep beneath the relative cool of the Akrotiri Peninsula, the atmosphere inside the Chania Facility was equally charged, though with anticipation rather than solar radiation. The newly arrived micro-servos, delivered at considerable risk by OPSEC lay gleaming under the sterile lights of Montane's lab. These tiny, radiation-hardened components were crucial for the next phase of experimentation: achieving finer control over the resonant field geometry at higher energy levels, hopefully mitigating the dangerous instabilities encountered during Test Run Alpha Bravo Niner. Belgrade, Ekota, and Montane had spent days refining the theoretical models and recalibrating the Core emitters, incorporating the potential offered by the new hardware.

"The simulations suggest these servos will allow us to dampen the fourth-order harmonic resonance by eighty-seven percent," Ekota explained, pointing to complex waveforms scrolling across a monitor. "It should prevent the spatial distortion flicker we observed previously, even at slightly increased power levels."

Sophia Solo, overseeing the power sequencing preparations, remained skeptical. "Increased power means increased thermal load. The primary heat exchanger is already operating at ninety-five percent capacity under baseline conditions. If we push it harder, then we risk a cascade failure."

"The entropy reduction effect scales non-linearly with field intensity, Sophia," Belgrade countered, her voice betraying a rare edge of impatience. "We need to reach that higher threshold to achieve sustained cooling over a viable volume. The servos give us the control; we must accept the calculated risk on the power systems." Calculated risk. The phrase hung in the air. Sophia knew that calculated risk often meant a desperate gamble when dealing with theorists pushing the boundaries.

"All systems prepped," Montane reported calmly, her hands moving precisely over the control console. "Emitter alignment verified. Servo feedback loop integrated. Ready for initiation sequence on your command, Dr. Belgrade."

Belgrade hesitated for only a moment, glancing at the security feed monitor showing Nathan's stoic face in the command center. The external pressures were mounting. Nathan had made it clear their window of undisturbed operation might be closing. They needed results, tangible proof to justify OPSEC's continued protection and resource allocation. "Initiate sequence," Belgrade ordered, her voice calm but firm. The familiar resonant hum began to build within the Core chamber, deeper this time, underpinned by the almost imperceptible whine of the new micro-servos making infinitesimal adjustments. This time, they aimed not just for entropy reduction, but for a measurable, sustained temperature drop within the one-meter containment sphere–the first true demonstration of localized cooling. The cold against the heat.

Inside the Chania Facility, the resonant hum of the Core intensified. On the main display, the entropy reading within the containment sphere began to drop steadily, more smoothly than ever before. Then, another readout flickered to life: the internal temperature sensor. It dipped. One degree Celsius below ambient. Two degrees. Five. Ten degrees. A sustained, localized cold spot, carved out of the background heat by the precisely controlled resonant field. A collective gasp went through the control room. Montane allowed herself the barest hint of a smile. Ekota pumped a fist silently. Belgrade watched, mesmerized, tears welling in her tired eyes. It was working.

But Sophia Solo's voice cut through the elation. "Reactor temperature climbing past threshold! Heat exchanger output spiking! We're overloading!" Alarms began to shriek. The triumph was instantaneous, the potential consequence immediate. The cold they had generated came at the cost of immense heat elsewhere in the system—thermodynamics in action.

Simultaneously, deep within the rock, a sensitive seismic sensor, placed far from the Core chamber, registered a minute, anomalous tremor–not geological, but resonant, linked to the Core's operation at this new intensity. Miles away, a specialized sensor drone detected a faint, localized thermal *negative* anomaly near the surface above the old NATO base–a cold spot where there should only be sunbaked rock. It was fleeting, quickly masked by the facility's countermeasures, but it was detected.

The heat of potential meltdown battled the controlled cold of success within the Chania Facility. Outside, the heat of the Mediterranean sun beat down, while the cold eyes of reconnaissance drones and approaching patrols began to focus on the secrets hidden beneath the Cretan soil. The fragile balance was tipping.

CHAPTER 9

The klaxons' shriek echoed the frantic pulse hammering in Sophia Solo's temples. Red emergency lights painted the Chania Facility's control room in stark, hellish hues, reflecting off the sweat beading on her brow. The triumphant glow from the successful entropy reversal test had vanished, replaced by the immediate, visceral terror of imminent system failure.

Temperature readouts for the salvaged reactor core climbed relentlessly, pushing far beyond established safety margins. The primary heat exchanger, tasked with bleeding off the colossal waste energy generated by the Resonant Field Core, was screaming under the strain, its efficiency plummeting as thermal saturation kicked in. They had cooled a one-meter sphere by ten degrees, a monumental achievement, but the cost was threatening to melt down their entire sanctuary.

"Stabilize the reactor! Initiate emergency coolant flush!" Sophia snapped orders into the comms, her voice tight but controlled. Beside her, Belgrade looked aghast, the theoretical beauty of this success crashing against the brutal physics of energy conservation. Ekota frantically worked auxiliary controls, trying to reroute power and isolate failing systems, while Montane methodically began the shutdown sequence for the Resonant Field Core, coaxing the dangerous energies back into quiescence. The resonant hum subsided, replaced by the high-pitched whine of overburdened pumps and the hiss of emergency coolant flooding the reactor loop. Slowly, agonizingly, the temperature climb arrested, hovering just below catastrophic failure levels.

Silence fell, thick and heavy, broken only by the panting breaths of the team and the steady thrum of the now-strained life support. "Damage report?" Nathan's voice crackled over the comm from the central command hub, calm as ever, though Sophia detected a slight edge.

"Core stable, just," Sophia reported, wiping sweat from her eyes. "Primary heat exchanger is compromised. We pushed it too far, too fast. We can't run the Resonant Field at that intensity again without risking meltdown, not unless we significantly upgrade our heat dissipation capacity." Upgrading wasn't feasible; they barely had the parts to maintain the current systems. They were stuck. Their greatest success had simultaneously revealed a hard limit; a wall built of waste heat.

Belgrade paced the control room, running a hand through her already disheveled hair. "There must be a way. Perhaps refine the field harmonics further, reduce the energy conversion inefficiency?"

"The inefficiency is inherent!" Sophia shot back, frustration boiling over. "The Second Law doesn't yield ground willingly. We forced localized order, and the universe demanded payment in diffuse heat, right here in our laps. More power; more control... it means more waste heat. Basic thermodynamics." She turned to the facility schematics displayed on a large monitor, zooming in on the deeper levels. "OPSEC surveys were incomplete when they repurposed this place. There were auxiliary systems mentioned in the original NATO decommissioning reports – backup coolant loops, perhaps secondary heat sinks– located in Level Delta. It was sealed off, deemed unnecessary or too compromised."

"Level Delta is off-limits, Sophia," Nathan's voice was firm. "OPSEC protocols designate it Hazard Zone Four. Structurally unsound in parts, potential residual contamination from ordnance storage, incomplete environmental data. It wasn't secured."

"We don't have a choice, Nathan!" Sophia countered, her voice rising. "Without additional cooling, the Chania Directive hits a dead end right here and now. We can make micro-cold spots, but we can't scale up to anything meaningful without cooking ourselves. Level Delta might hold decommissioned industrial chillers, access to deeper geothermal vents, *something*. We have to look."

A long pause stretched over the comms. Sophia could picture Nathan weighing the risks: compromising security protocols, entering an unsecured zone, potentially exposing the team to unknown hazards versus the certain failure of their primary mission. The red emergency lights still pulsed, casting long shadows. "Very well," Nathan finally conceded. "Minimal team. You, me, one tech. Full hazmat suits, standard Hazard Zone protocols. We reconnoiter Level Delta, assess potential auxiliary systems. No unnecessary risks."

Sophia nodded, relief warring with trepidation. She selected David, a quiet, brilliant technician with experience in older hydraulic and atmospheric systems, to accompany them. Preparations were swift. Donning the bulky hazmat suits felt like sealing themselves into coffins. The airlock leading to Level Delta hissed open, revealing not a modern tunnel, but an older, rougher excavation, the air thick with the scent of damp concrete, stale machine oil, and decades of undisturbed dust. Emergency lighting flickered intermittently, casting eerie shadows down the corridor.

They moved cautiously, their helmet lamps cutting beams through the oppressive darkness. The architecture was different here–starker, more utilitarian, clearly from the Cold War era—1947-1991. Thick steel blast doors stood ajar, revealing empty storage bays or stripped equipment rooms. Degraded NATO insignia, faded warnings in multiple languages, and cryptic alphanumeric codes stenciled on walls hinted at the level's former purpose. Sophia consulted a damaged schematic downloaded from OPSEC's patchy archive, trying to match the layout to their intended path towards the auxiliary engineering section.

"Schematic indicates a primary coolant reservoir access point should be through here," Sophia said, her voice slightly muffled by the suit's comms. She indicated a heavily reinforced door marked Sector 7G–Restricted Access. The locking mechanism was archaic, purely mechanical. David, deft despite the bulky gloves, worked on the heavy tumblers. With a groan of protesting metal, the door swung inwards. Beyond lay not a reservoir access tunnel, but something else entirely: a climate-controlled chamber, smaller than expected, lined with rows of magnetic tape reels, microfilm canisters, and surprisingly, a single, bulky computer terminal encased in protective housing, its screen dark.

"This isn't on the schematic," Nathan stated, his hand instinctively moving towards the sidearm holstered on his suit. He swept his lamp around the room. It felt... preserved. Deliberately sealed, not just abandoned. "OPSEC missed this during their initial sweep."

Sophia approached the terminal. It looked like a relic from the late 20th century, built like a bunker itself. "There might be power," she murmured, tracing sealed conduits leading into the wall. David examined the tape drives and microfilm readers. "Obsolete formats. We might not have the equipment to read these, even if they are not degraded."

But the terminal flickered. A low green light appeared on its casing. "Emergency power backup," Sophia breathed. "Still holding a charge after all these years." She carefully accessed a maintenance port, connecting a diagnostic tool from her kit. Lines of archaic code scrolled across her suit's integrated heads-up display. "Trying to bypass security… old protocols, but layered." Minutes passed in tense silence, broken only by the hiss of their suit recyclers. Then, the main screen sputtered to life, displaying a stark, blocky emblem Sophia didn't recognize, beneath it the words: 'NATO TOP SECRET–PROJECT CHIMERA-Special Projects Directorate—Archival Access'.

Chimera? Sophia exchanged a look with Nathan. This wasn't ordnance storage or standard engineering. This was something else. With David's help navigating the clunky interface, Sophia began accessing the archived files.

They were fragmented, some corrupted, protected by keyword locks they had to painstakingly circumvent. But slowly, a picture began to emerge. Project Chimera, active from the mid-1970s to the late 1990s, conducted right here in Level Delta. It wasn't focused on missiles or electro-magnetic (EM) conventional warfare. Its mandate, according to the recovered logs, was far stranger: investigating localized atmospheric phenomena and exploring the potential for manipulating electromagnetic fields using the unique tectonic-magnetic properties of the Akrotiri massif.

Logs detailed experiments involving generating powerful, complex EM fields deep within the rock, attempting to create localized ionospheric disturbances for communication disruption or, more ambitiously, environmental modification effects. There were references to unexpected energy fluctuations, difficulties in containing the fields, and anomalous sensor readings–descriptions chillingly similar to the instabilities Belgrade's team had encountered. Then, abruptly, the logs detailed a major incident in 1998: an uncontrolled energy cascade during a high-power field test, resulting in severe equipment damage, localized spatial distortions reported by startled technicians (dismissed as sensor ghosts in the official summary), and lingering EM field instabilities within the bedrock itself.

Shortly after, Project Chimera was officially terminated. Level Delta was partially decommissioned, key equipment removed or destroyed, and this entire section, including the archive room, was sealed off under layers of security and classification, effectively buried and forgotten even within NATO's labyrinthine bureaucracy. OPSEC, focusing on the more modern upper levels related to missile operations, had either missed this deeper archive or deemed the sealed level irrelevant or too hazardous to fully explore during their initial hurried setup of the Chania Facility.

"So, we aren't the first," Nathan murmured, his voice tight. "They were poking at something similar down here decades ago. And they lost control."

"It's more than that," Sophia said, pointing to a recovered geological survey map integrated into the Chimera files. "Look. Chimera wasn't just using the mountain; they were studying *why* this location was suitable. There's a unique convergence of subterranean fault lines and mineral deposits directly beneath us, creating anomalous geo-magnetic flux lines. It's what made their EM experiments possible, and likely what makes Belgrade's resonant fields interact so powerfully." She paused, tracing a highlighted area on the map. "And according to this survey, their 1998 incident may have created a residual fracture zone, an area of lingering field instability, deep beneath the level housing our Resonant Field Core."

The implications struck them like a physical force. The instabilities Belgrade's team were fighting weren't just random quantum noise or flaws in their theory. They might be interacting with a pre-existing, damaged energy field embedded in the very rock around them—a ghost left behind by Project Chimera. The spatial distortions Montane observed weren't necessarily a product of entropy reversal itself, but potentially a resonance with this older, unstable field. Their state-of-the-art experiment was inadvertently tickling a dragon that NATO had tried, and failed, to tame forty years earlier.

"Does this help with the cooling?" Nathan asked, pulling them back to the immediate crisis.

Sophia scanned the recovered technical logs. "Possibly. Chimera used a massive geothermal heat exchange system, tapping into a deep fault line. It was damaged in the 1998 incident and listed as decommissioned. But the primary conduits might still be accessible from this level. If we can tap into them, bypass the damaged sections…it could give us the heat dissipation capacity we need." It was a long shot, requiring navigating more of the hazardous Level Delta, but it was a tangible possibility.

"But it also means," David added quietly, "that the bedrock itself might be unstable. Powering up Belgrade's device could potentially re-energize whatever Chimera left behind, with unpredictable results."

They stood in the silent archive room, the weight of the past pressing down on them. They had descended seeking a solution to their present crisis and had instead unearthed a hidden history, one that intertwined dangerously with their future. The Chania Facility wasn't just a repurposed missile base; it was a place with its own secrets, its own potential energy waiting to be unleashed or dangerously provoked. The mission had just become infinitely more complex, the ground beneath their feet suddenly feeling far less solid. The heat outside continued its relentless assault on the planet, while the cold they sought to create was now entangled with the chilling legacy of a forgotten Cold War experiment.

CHAPTER 10

Deep beneath Crete, in the stale, artificially lit confines of Level Delta, the air in Sophia Solo's environmental suit tasted recycled and faintly metallic. Following the damaged schematic and the recovered Chimera logs, she, Nathan, and David navigated deeper into the hazardous zone. The corridors grew rougher, transitioning from poured concrete to sections blasted directly from the rock, walls glistening with moisture and streaked with mineral deposits. Faint electromagnetic fields, likely remnants of the 1998 incident, occasionally caused static bursts in their comms or flickers on their suit displays–unsettling reminders of the invisible forces lurking within the mountain.

They located the area designated as the primary access point for the geothermal heat exchange system. It wasn't a room, but a massive cavern, partially collapsed at one end. Dominating the space were the remains of huge, heavily corroded pipes, valves the size of small vehicles, and what looked like the base of an enormous turbine, now shattered and silent. "This is it," Sophia breathed, awe mixing with dismay. "The scale… it's far larger than the logs implied." It was clear Project Chimera had been operating with energy levels far exceeding what Belgrade's team currently handled.

Nathan scanned the cavern with thermal imagers and radiation detectors. "Radiation levels are slightly elevated near the collapsed section, likely fractured rock exposing deeper mineral veins. Otherwise, stable. But the structural integrity here is questionable." He pointed towards the far wall where the main geothermal conduit should penetrate deeper into the fault line. "That's our target. If we can access that conduit, bypass the damaged turbine assembly and reservoir…"

David moved towards the conduit access panel, examining the heavy seals. "These look intact. Standard deep-earth pressure fittings. But look at this." He indicated faint scorch marks and warped metal around the panel's edges. "Something overloaded here. The 1998 incident?" Sophia checked the Chimera logs again. The 1998 cascade had originated near the field generator, but the energy surge had propagated throughout their entire system, including the geothermal loop. "They tried to vent excess energy through the geothermal exchange," she realized. "It wasn't designed for that kind of load."

As David began carefully working on the access panel's bolts, Sophia and Nathan examined the surrounding area more closely. Near the base of the shattered turbine, half-buried in debris, was something Nathan's initial sweep had missed: the skeletal remains of a human body, still clad in the remnants of a heavy-duty protective suit similar to their own, but of an older design. The suit was fused and melted around the body in places, suggesting exposure to extreme heat or energy. There was no identification visible. "Not listed in the official Chimera incident report," Nathan stated grimly, kneeling beside the remains. The report mentioned equipment damage and minor injuries, but no fatalities. Another truth buried in the ruins–the official story was a sanitized lie. Project Chimera hadn't just failed; it had killed, and then covered it up.

David finally managed to release the last bolt on the geothermal conduit access panel. With a groan of tortured metal, they eased the heavy plate aside. Instead of a simple pipe, it revealed a complex junction, insulated conduits branching off into the rock. More importantly, mounted within the junction housing was a heavily shielded diagnostic unit, humming faintly with residual power. Sophia accessed its interface. It contained logs from Project Chimera's final moments–detailed sensor readings from the 1998 energy cascade.

The data was startling. It showed the geothermal exchange hadn't just been overloaded; the energy surge, interacting with the anomalous geo-magnetic fields Chimera had been manipulating, had caused a feedback loop. The system hadn't just vented energy; it had briefly *amplified* it, creating a localized rift or instability in the deep bedrock–the fracture zone the survey map hinted at. This rift wasn't just passively unstable; according to the final frantic log entries before the system failed, it exhibited properties suggesting it could absorb *and* release tremendous amounts of energy unpredictably if stimulated by the right resonant frequencies.

"Oh my God," Sophia whispered, stepping back from the console. "It's not just a ghost field down here. It's a wound. A wound in the bedrock that can bleed energy. The 1998 incident didn't just damage the cooling system; it created a potential amplifier for the very forces Belgrade is trying to control." The truth buried in these ruins wasn't just about past failures or hidden dangers. It was that the Chania Facility was built atop a geological time bomb, one inextricably linked to the physics of their own experiment. Trying to fix the heat problem might just ignite a far greater fire.

CHAPTER 11

The pulsing red emergency lights in the Chania Facility's control room had subsided, replaced by the harsh, steady glare of standard illumination, but the tension remained thick enough to taste. They had averted meltdown, but the system was critically damaged. Running the Resonant Field Core at the intensity needed for meaningful cooling was now definitively impossible without accessing Level Delta's hazardous, half-understood geothermal systems.

"We bought ourselves time, nothing more," Sophia stated flatly, stripping off her diagnostic gloves. "The exchanger won't handle another run like the last one. Any attempt to reach sustained cooling requires tapping into whatever Chimera left below. And we now know that carries… significant risks." She glanced towards Belgrade, who looked pale and profoundly shaken, the revelation from Level Delta weighing heavily upon her. Her elegant theories had collided head-on with a dangerous, hidden history etched into the very rock surrounding them.

Nathan, ever the pragmatist, focused on the immediate external threats. "OPSEC confirms private security contractors are enroute, likely disguised civilian vessel, ETA within twelve hours. Concurrently, a shore party is making landfall Sector North-East. Four specialists, moving cautiously." He projected a map onto the main screen, highlighting the team's position relative to the facility's outer sensor perimeter. "They're probing, acting on intel; they don't know what they're looking for yet, but they're getting closer."

"Can we jam their sensors? Divert them?" Ekota asked, his voice strained. He was still rattled by the near-meltdown and the subsequent revelations about Project Chimera. The idea of external hostiles closing in felt like another layer added to an already unbearable pressure.

"Standard EM countermeasures are active," Nathan replied. "But the shore party includes veterans familiar with this terrain and likely equipped with hardened gear resistant to basic jamming. The security contractors will have state-of-the-art intrusion tech. We rely on stealth, misdirection, and minimal engagement. My teams are deploying seismic decoys near the shore party's approach vector, trying to lure them towards the old surface bunkers."

Deep within Level Delta, David worked under portable lighting, examining the corroded access panel to the geothermal conduit junction. The air remained thick with the scent of ozone and damp rock. "The seals are brittle, Nathan," David reported over the comm, his voice muffled by the environmental suit. "Opening this will be... delicate. And the diagnostic unit Sophia accessed? Its residual power signature is fading. Whatever final logs it held about the 1998 incident, we might lose them if we don't extract the data module soon." The past wasn't just haunting them; it was actively decaying.

The shore party moved with practiced silence through the thorny scrub covering Akrotiri's rocky slopes. Led by a grizzled veteran who knew these hills from his days patrolling against smugglers, they swept the area with handheld sensors. They found signs of recent passage– faint boot prints near an overgrown access road and a discarded nutrient bar wrapper. Then one of their sensors picked up faint, rhythmic seismic vibrations originating further inland. The vibrations seemed to originate near the main NAMFI complex, but deeper, perhaps underground. This was more than smuggler activity.

Nathan watched the intruders' hesitation on his tactical display. The seismic decoys were working, drawing their attention. He ordered his perimeter team– two highly trained operatives moving silently through ventilation shafts and natural crevices–to observe, not engage, unless directly compromised. Every encounter increased the risk of discovery, of leaving evidence. The goal was deterrence through ambiguity, making the intruders believe the area was either naturally unstable or haunted by unreliable old tech, not actively occupied. Suddenly, a different alarm pinged on Nathan's console. A proximity alert from an underwater sensor near the facility's hidden submersible access point. "Unidentified submersible drone detected, small profile, attempting passive reconnaissance of access tunnel," JoAnn's voice came back, strained. "A new player? Deploying countermeasures." The new arrival wasn't the main vessel, still hours out, but likely an advance scout deployed from a smaller, faster boat. The multi-pronged assault had begun.

Inside the main lab, Belgrade, Montane, and Ekota tried to refocus on the science, specifically the implications of the Chimera data. "If the 1998 incident created a stable energy rift or fracture zone," Belgrade mused, sketching rapidly on her whiteboard, "it might act as an uncontrolled resonant cavity. Our field could be exciting it, causing the instabilities. But... could it also be harnessed? Could we use it as part of the energy dissipation mechanism? Channel the waste heat *into* the rift instead of overwhelming our exchangers?"

Montane looked skeptical. "Harnessing an uncontrolled energy cascade based on forty-year-old fragmented data? The risks are astronomical. We barely understand the rift's properties. It could amplify our field uncontrollably, or destabilize catastrophically." She pointed towards the recovered Chimera

54

diagnostic unit data Sophia streamed from Level Delta. "We need that full data set before even considering such a path."

Sophia's voice cut in, breathless. "Nathan, David's through the primary seal on the geothermal conduit access. Inside... it's not just pipes. It looks like a secondary control node for the Chimera field generator itself. Heavy shielding, but damaged. And there's definitely residual energy readings." The connection between the failed cooling system and the dangerous legacy of Project Chimera was direct, physical. Accessing the cooling solution meant confronting the source of the instability head-on.

Nathan saw the intruders change vector, bypassing the decoys, heading directly towards a vulnerable sensor node disguised within the fence line. "Perimeter team, Intercept Vector Gamma. Non-lethal takedown authorized if necessary. Maintain absolute silence. No comms chatter." The two OPSEC operatives melted from the shadows, equipped with sonic stunners and capture foam launchers, converging silently on the approaching intruders. The first real confrontation was imminent.

Simultaneously, JoAnn reported success neutralizing a recon drone near the submersible access. "Drone disabled via focused EMP burst. Minimal collateral EM signature detected. But they know we're actively defending now. Expect escalated response from the main vessel upon arrival." The brief victory only confirmed the escalating danger.

The battle for survival was unfolding on multiple fronts, a complex, chaotic dance of stealth, technology, desperation, and violence. Inside the Chania Facility, the team wrestled with damaged systems and the dangerous legacy buried beneath them. Outside, Nathan's hidden defenders moved to intercept the intruders. Underwater, OPSEC forces neutralized one probe while bracing for the main force arrival. Each action, each decision, rippled outwards, tightening the net around the Chania Facility, threatening to expose the secret keepers and their world- altering project to the brutal realities of Earth 2050. The fragile threads of hope were strained to the breaking point, threatening to snap under the combined pressure of heat, cold, and human conflict.

CHAPTER 12

Deep underground, Nathan watched the tactical display confirm the successful, non-lethal neutralization of the shore party. "Sweep team confirms four hostiles secured, sedated, prepped for extraction," the encrypted text scrolled across the screen. "Minimal physical evidence left at intercept point, but their comms likely logged sensor anomalies before takedown. Recommend immediate relocation of secured personnel via OPSEC sub-surface transport." Nathan acknowledged, initiating the complex extraction protocol. They had neutralized the immediate threat, but the intruders had detected anomalies. Their failure to report back would trigger a larger response eventually. The clock was ticking faster now. More importantly, the very act of defending their perimeter confirmed something *was* there to defend.

The pressure translated downwards, into the humid, tense air of the main control room. Sophia Solo, back from her initial foray into Level Delta, presented her findings starkly. "The primary heat exchanger is shot. We cannot repeat the high-intensity test without risking core breach. Level Delta offers a potential solution–the old Chimera geothermal conduit. But accessing it means potentially interacting with the unstable energy field their 1998 incident left behind. Furthermore," she indicated the decaying diagnostic unit's interface still displayed on a side monitor, "the unit containing the full logs of that incident is losing power. We need that data to understand the risks, perhaps even model a safe interaction."

Belgrade, who had been staring intently at the recovered Chimera schematic fragments, looked up: "The instability… perhaps it's not just a risk, but a component. If the 1998 incident created a stable rift capable of absorbing or channeling energy, could we modulate our resonant field to *use* it as a heat sink? Direct the waste energy into this… this geological wound?"

"You want to deliberately interface with an uncontrolled energy cascade based on forty-year-old fragmented logs and a dying diagnostic unit?" Montane asked, her usual calm ruffled. "The potential for catastrophic feedback, for amplifying the very instabilities we observed in Alpha Bravo Niner, is enormous. We could destabilize the entire bedrock."

"We are out of options!" Belgrade countered, slamming her hand on the console. "We achieved sustained cooling! Proof of concept! But it's useless if we cannot scale it, if we cannot manage the thermal load. OPSEC protection is contingent on progress. We need a breakthrough, not incremental safety." She turned to Nathan, whose image remained impassive on the comm screen. "Nathan, we need that diagnostic data, and we need access to the geothermal conduit. Sophia must go back to help David in Level Delta."

Nathan considered, the silence stretching. The risks were immense–sending personnel back into the hazardous, unstable Level Delta while external threats mounted. But Belgrade was right; stagnation meant discovery and failure. "Authorized," Nathan replied finally. "Sophia proceed with data extraction and conduit assessment. Full hazmat suits, maintain constant comms link. My primary focus remains external defense. You are on your own down there regarding structural or environmental hazards." Sophia nodded, already pulling schematics for the diagnostic unit's data module extraction.

Offshore, the sea seemed unnaturally calm as the Endevor, a sleek, repurposed oceanographic research vessel flying the neutral colors of a defunct marine science consortium, dropped anchor several kilometers northeast of Chania. Onboard, however, activity hummed with quiet, corporate efficiency. The intruder command, monitoring remotely from Geneva, watched through the eyes of multiple sophisticated drones deploying from the ship–aerial units equipped with high-resolution multi-spectral imagers, and submersible drones armed with passive-sonar arrays and deployable micro-sensors designed to map underwater energy fields and potential access points. Now, they needed confirmation and characterization before deploying the main intrusion team.

Within the echoing cavern of Level Delta, Sophia and David worked with intense focus under portable lamps. The diagnostic unit containing the final Chimera logs was proving stubborn. Its housing, designed to withstand extreme conditions, required specialized tools to open without damaging the fragile data module within. David, his faceplate beaded with condensation, carefully applied a micro-drill to the casing screws. Sophia monitored the unit's fading power signature. "Thirty minutes of residual power, maybe less," she reported grimly. "If we don't get this module out soon, the data degrades completely."

As David worked, Sophia examined the main geothermal conduit junction. The warped metal and scorch marks around the access panel were stark evidence of the forces unleashed in 1998. Using a portable scanner, she probed the rock beyond the conduit. Her display flickered erratically. "Nathan, getting strong, unstable EM readings emanating from *within* the fault line access. It's not just residual; it feels active. Resonating faintly."

"Correlates with Belgrade's hypothesis?" Nathan asked.

"Potentially. Or it correlates with something about to go very wrong if we disturb it further," Sophia replied, her voice tight. The choice loomed: risk extracting the crucial data and potentially accessing the cooling system, or retreat from a danger they were only beginning to comprehend.

Suddenly, a new alarm blared through Nathan's command hub, overriding the tactical displays. An automated security alert from the deepest sensor network. "Intrusion detected! Level Gamma, Ventilation Shaft Access Point Seven! Multiple pressure seals bypassed remotely!" Level Gamma was directly above Sophia and David in Level Delta, dangerously close to the main facility core. The intruders were not just probing; they were inside. An infiltration team, likely deployed from one of the submersible drones, had found a vulnerability, bypassing the outer defenses electronically.

"Sophia, David, hostile intrusion Level Gamma! Abort data extraction, secure your position, await tactical support!" Nathan ordered, simultaneously dispatching his internal security team towards the breach point.

But Sophia shook her head, looking at the diagnostic unit's fading power light. "Negative, Nathan! Almost there! This data could be critical to understanding the rift itself!" David applied a final torque, and the data module casing popped open.

On the bridge of the Endevor, the mission received confirmation: Intrusion team Pathfinder successfully bypassed primary seals in Ventilation Shaft Seven. They were moving towards suspected core facility levels. Simultaneously, sensor drones reported faint, unusual energy fluctuations deep within the peninsula– consistent with the diagnostic unit Sophia was accessing, and perhaps, the stirring instability of the Chimera rift responding to the nearby resonant fields.

Nathan saw the intrusion alert confirmed on his display, saw the location– Level Gamma, directly adjacent to the Resonant Field Core chamber. He knew the Endevor was deploying more assets. He heard Sophia reporting unstable energy readings from the Chimera rift below. The carefully maintained secrecy, the layers of defense, were collapsing simultaneously from multiple directions. Stealth was failing. Misdirection was failing.

He looked at the status board for the Resonant Field Core–offline but potentially ready. He looked at the data stream flickering from Level Delta, hinting at both immense danger and potential control. A decision point, brutal and immediate, arrived. They could lockdown, attempt to fight a losing defensive battle against superior forces, hoping OPSEC intervened. Or they could force the issue. They could turn the liability into a weapon.

"Belgrade!" Nathan's voice crackled with uncharacteristic urgency. "Intruders are inside Level Gamma! More backup imminent! Can you initiate a controlled, localized energy pulse using the Core, interfaced with the Level Delta conduit? Something... deterrent?"

Belgrade stared, aghast. "Interface with the rift? Based on partial data? Intentionally? Nathan, the risks..."

"The alternative is capture or destruction within the hour!" Nathan cut her off. "Give me something to push them back!"

Sophia's voice came over the comm, breathless, holding up the small, extracted data module from the Chimera diagnostic unit. "Got it, Nathan! Got the data! But the EM fields down here... they're spiking! Something's reacting to the Core's residual signature, or maybe... maybe to the intrusion upstairs?"

This was it. The confluence. The moment where desperation, discovery, and danger converged. The turning point wasn't a choice made in calm reflection, but one forged in the crucible of imminent destruction. Belgrade looked at Montane, at Ekota, then back at the Core status board. She took a deep breath. "We can try. A narrow-beam resonant pulse, channeled through the geothermal conduit access... aiming for maximum EM interference, minimal entropy effect. Theoretical. Untested. Highly unstable." She met Nathan's gaze on the monitor. "Initiate on your command." The battle for survival had just become a radical, terrifying experiment.

CHAPTER 13

The command, when Belgrade finally gave it, "Execute pulse, narrow beam, geothermal conduit interface" hung in the super-saturated tension of the Chania Facility control room for less than a heartbeat before Montane's fingers flew across the console. Power surged, not with the full brute force of the earlier entropy reversal test, but with a precisely modulated intensity, directed downwards, towards the newly accessed junction in Level Delta, towards the fifty-two-year-old rift in the bedrock left by Project Chimera.

Deep below, in the echoing cavern where Sophia and David huddled near the conduit access panel, the air crackled. Residual lights flickered violently. A low, bone-jarring hum resonated through the rock itself, far deeper than the Resonant Field Core's usual signature. The unstable EM readings Sophia had detected went off the scale, her scanner screaming electronic protest before overloading. Through the thick viewport of her helmet, she saw faint, ethereal blue traceries of light dance along the corroded metal of the conduit junction–St. Elmo's fire on steroids, generated by immense, localized field potentials. The Chimera rift was awake.

Simultaneously, in Level Gamma, the intrusion team, designated "Pathfinder," experienced chaos. Moving stealthily through Ventilation Shaft Seven towards the suspected core facility, their sophisticated electronic gear suddenly went haywire. Heads-up displays dissolved into static, encrypted comms filled with deafening white noise, targeting systems failed, and navigational aids spun uselessly. The leader felt his cybernetic eye implant short out with a painful jolt. They were blind, deaf, and technologically crippled, caught deep within a hostile environment that had just demonstrated an offensive capability far beyond conventional electronic warfare.

Outside, aboard the Endevor, the deployed drones monitoring Akrotiri simultaneously reported catastrophic system failures. Aerial units lost navigational lock and telemetry, tumbling from the sky or automatically initiating emergency return protocols with corrupted data logs. Submersible units experienced similar electronic disruption, their passive-sonar arrays overwhelmed by a burst of complex EM noise that defied classification. Whatever was happening at the derelict NATO base wasn't just producing energy; it was weaponizing it in a way the intruders' threat matrix hadn't anticipated.

Back in the Chania Facility control room, the pulse lasted only three seconds. Montane cut the power feed precisely. The deep hum from below subsided, leaving only the strained whine of the overburdened facility systems. Alarms shrieked anew as secondary power relays, stressed by the directed energy surge, tripped offline. Ekota scrambled to restore power to essential systems. Belgrade stared at the sensor readouts, aghast. "The efficiency... the energy conversion into directed EM was far higher than predicted. The rift *amplified* the pulse, focused it."

Sophia's voice came over the comms, shaky but clear, from Level Delta. "Confirmed amplification. Residual EM fields down here are... chaotic but settling. The diagnostic module is intact, extracted. But the conduit junction housing shows signs of extreme thermal stress, verging on meltdown. We bought ourselves a deterrent, maybe, but we may have permanently damaged the only viable interface with the geothermal cooling system." She paused, her breath ragged. "...the final Chimera logs on this module...they reference 'exotic particle generation' during the 1998 cascade. Unidentified, uncontained. They didn't just lose control; they might have created something they couldn't even measure."

The weight of this revelation settled over the control room. They had used the Chimera rift to deter the immediate intruders. But they had done so blindly, provoking forces they didn't understand, based on fragmented warnings from a dead past. They might have traded immediate survival for a far more dangerous long-term instability. The very bedrock beneath them could now be seeded with exotic particles or unstable energy potentials, a permanent, invisible legacy of their desperate act.

Nathan absorbed the information rapidly. "Pathfinder team is electronically isolated, likely retreating if they can navigate manually. Their drones neutralized. We have a window, perhaps hours, before they regroup or they escalate." He turned his attention back to the internal crisis. "Sophia, assess conduit integrity. Belgrade, Montane, Ekota – analyze the pulse effects, recalibrate models based on the observed amplification. We need to understand what we just did, and if it can be controlled, repeated, or if we simply have to seal Level Delta forever and find another way."

The atmosphere within the facility shifted subtly. The immediate panic subsided, replaced by a daunting awareness of the new reality. They had demonstrated power, undeniable and terrifying. The secret of their existence was likely compromised beyond plausible deniability. But the nature of that power, its connection to the unstable Chimera legacy, and their ability to control it remained terrifyingly uncertain. They had crossed a threshold. The mission was no longer just about reversing entropy; it was now inextricably linked to managing the dangerous forces sleeping within the mountain.

JoAnn Solo, monitoring the chaos remotely felt a similar shift. The EM pulse, though localized near Crete, had been detected by sensitive OPSEC monitoring stations across the Eastern Mediterranean. It was orders of magnitude stronger than anything the Chania Facility should have been capable of producing, confirming anomalous amplification. Combined with the intruders' activity, it signaled a catastrophic failure of containment. Her orders remained 'Evade and Observe', but JoAnn knew OPSEC leadership would now be forced into agonizing reappraisals. Was the Chania Directive still viable? Was it too dangerous? Could it be controlled, or should it be terminated? The hidden network she served felt poised on the brink of extermination.

For Geneva command and control, the reports filtering back from the Endevor were electrifying. Catastrophic equipment failure, massive EM pulse, confirmation of offensive capabilities far exceeding projections–this wasn't some minor research project. This was paradigm-shifting technology, likely energy or weapons-related, potent enough to disable the best equipment. They initiated Phase Three protocols: deployment of the main contractor assault force, heavily shielded against EM effects, equipped for breaching hardened facilities. Objective: secure the technology and its creators at all costs. Capture, contain, control. The time for reconnaissance was over. Akrotiri held the key to future power, and they intended to possess it.

The pulse, intended as a desperate deterrent, had instead acted as a global announcement, albeit one spoken in the language of energy signatures and system failures, decipherable only by those with the means to listen and understand. It confirmed the existence of something extraordinary beneath Crete, shattering the fragile secrecy that had protected Belgrade's team. It forced every player to recalculate, to reimagine the strategic landscape.

Inside the Chania Facility, the team faced their own reimagining. They were no longer just scientists pursuing a theoretical solution to global warming. They were custodians of a dangerous legacy, manipulators of forces they barely comprehended, besieged by powerful enemies, and standing atop a geological wound that might be key to salvation or the trigger for annihilation. The successful cooling test felt like a lifetime ago. The immediate battle for survival had been won, momentarily, but it had irrevocably changed the nature of their quest. The path forward led deeper into the mountain's secrets, towards confrontation, and into a future where the power to remake the world felt terrifyingly close, and terrifyingly uncontrollable. The world hadn't ended in fire or ice yet, but the potential for both, radiating from their subterranean sanctuary, had just become undeniable.

CHAPTER 14

The immediate aftermath of the energy pulse hung heavy over the Chania Facility, a silence more profound than the preceding chaos. Alarms were silenced, secondary systems slowly stabilizing under Ekota's frantic ministrations, but the air itself felt charged, expectant. The brief, violent demonstration of power had repelled the initial probes, but everyone knew it was a Pyrrhic victory, bought at the cost of anonymity and potentially system integrity. They had shouted their existence to a world ill-equipped to understand, inviting scrutiny from forces far more dangerous than curious onlookers.

Nathan watched the tactical displays with grim intensity. The intruders' recon drones were offline, either destroyed or withdrawn. The shore party, neutralized and extracted via OPSEC sub-surface transport just minutes earlier, represented a temporary success but a long-term complication. Their failure to report back, coupled with the detected energy signatures, would inevitably trigger a larger response. The clock wasn't just ticking; it was accelerating towards midnight. His internal security teams were on high alert, reinforcing potential breach points identified during the Pathfinder intrusion, but Nathan knew their small force couldn't withstand a determined assault.

In the main lab, the mood was a volatile mix of scientific excitement and existential dread. Sophia Solo, still processing the data recovered from the Chimera diagnostic module, worked alongside Belgrade and Montane. The final logs confirmed their worst fears and wildest hopes simultaneously. Project Chimera's 1998 incident hadn't just damaged the geothermal conduit; it had created a persistent, localized spatial distortion—a micro-rift—within the bedrock, capable of interacting with focused energy fields in highly unpredictable ways. It possessed properties that could potentially absorb immense thermal energy, but also properties that suggested catastrophic instability if improperly stimulated.

"The Chimera team called it the 'Pauli Anomaly'," Sophia reported, pointing to a complex diagram recovered from the module. "After the Austrian-born physicist Wolfgang Pauli who made major contributions to quantum mechanics, quantum field theory, and solid-state physics, and successfully hypothesized the existence of the neutrino. Because its energy absorption properties seemed to make energy vanish from local spacetime. But their final logs are frantic, incomplete. They mention uncontrolled particle decay, temporal fluctuations near the anomaly's core during the cascade. They didn't understand it. They sealed it off because they were terrified of it."

Belgrade, however, saw potential where Sophia saw peril. "Forgetfulness... or transference? If the anomaly acts as a stable conduit to a higher dimension, or even just dissipates energy into the quantum field more efficiently... it could be the heat sink we need! We modulate our field to resonate *with* the anomaly, not against it. We use it."

"Based on fragmented logs from a failed Cold War project that likely killed some of its researchers?" Montane countered, her voice sharp. "We amplified our pulse using it unintentionally and nearly tore Level Delta apart. Deliberate interaction is reckless. We need weeks, months, of simulations, assuming the data module is even complete and accurate."

"We don't have weeks!" Belgrade retorted. "We have hours. Nathan, what's the status of the geothermal conduit junction?"

"Sophia's initial assessment stands," came the reply over the comm. "Severely compromised by the pulse amplification. Accessing it for cooling purposes would require extensive repairs in Level Delta, exposing personnel for prolonged periods. Attempting to use it to deliberately channel energy towards the Pauli Anomaly is currently impossible due to the damage." The desperate gambit had obliterated the very path Belgrade envisioned.

Aboard the Endevor, the crew coordinated the preparations for a Phase Three assault. Heavy transport drones began lifting off the deck, carrying heavily armed and shielded corporate contractors–veterans of resource wars and corporate extraction operations from across the globe. Smaller, nimbler breaching drones were programmed with entry points derived from Pathfinder's partial reconnaissance before they were disabled. The EM pulse confirmed a high-value target, likely energy technology far exceeding current capabilities. Capturing it would be a major coup, cementing dominance in the fractured European theatre. The orders were clear: secure the source, retrieve the technology, eliminate indigenous resistance. Standard operating procedure.

JoAnn Solo felt the OPSEC network contract around her. Control's communications became clipped, focused solely on situational awareness near Crete. Her request for counter-drone assets was denied. Intel chatter suggested fierce internal debate within OPSEC leadership: cut their losses, classifying the Chania Facility as compromised and unsalvageable? Attempt a high-risk extraction of Belgrade and the core team, abandoning the facility? Or gamble on providing limited, deniable support? The ambiguity was agonizing. JoAnn continued her surveillance of the Endevor, feeding targeting data into OPSEC databases, hoping someone, somewhere, would authorize action before intruders overwhelmed the facility.

Inside Level Delta, Sophia and David carefully documented the state of the geothermal conduit junction and the surrounding area, focusing on the damaged Chimera control node. The final diagnostic logs from the extracted module were now being analyzed by Ekota back in the main control room. His initial findings were disturbing. The logs confirmed Chimera's accidental creation of the Pauli Anomaly, but also detailed their frantic, failed attempts to stabilize or neutralize it. Their conclusion, buried in corrupted data fragments, suggested the anomaly possessed a form of memory—it reacted more strongly, more unpredictably, to energy patterns similar to those that created it. Belgrade's resonant field, while different in kind, shared fundamental characteristics with Chimera's EM fields. Each activation of the Resonant Field Core, especially the recent high-intensity pulse, might be teaching the anomaly, potentially lowering its threshold for uncontrolled energy release.

"It means Belgrade's idea of using the rift as a heat sink is potentially suicidal," Ekota reported, his voice strained. "Deliberately resonating with it could trigger the same kind of cascade that destroyed Chimera, possibly worse now that it's been primed by our own activities." The potential solution to their cooling problem was also, quite possibly, the trigger for their annihilation. The mountain wasn't just damaged; it was learning, remembering, waiting.

Nathan processed this new information alongside the tactical updates. Intruder drones were re-establishing aerial surveillance, more cautiously this time. Their blockade was tightening. Time was running out. The path through Level Delta seemed fraught with potentially catastrophic danger. Defense against imminent assault seemed unlikely to succeed. He looked at the facility status board, at the Resonant Field Core, currently offline but charged. He looked at the simplified model Ekota had generated of the Pauli Anomaly's potential energy states based on the Chimera data. A desperate, almost unthinkable strategy began to form in his mind, born not of science but of pure tactical necessity.

The echoes of war were reverberating through the ancient landscape. Not just the echoes of the recent EM pulse and the scrambling responses, but the deeper echoes of the Cold War's secret projects, the echoes of collapsing global structures, the echoes of humanity's long, destructive relationship with power. These echoes converged on Crete, focusing on the small group of scientists and soldiers trapped between a hostile world outside and a potentially cataclysmic force stirring beneath their feet. The turning point had passed; the endgame, in fire, cold, or some unimaginable synthesis of the two, was rapidly approaching.

CHAPTER 15

The air crackled, thick with the ozone tang of unleashed energy and the bitter scent of desperation. Nathan watched the tactical display, a constellation of hostile indicators converging on their subterranean sanctuary. The intruders' main assault force was closing, shielded against conventional EM attacks, their breaching drones probing for weaknesses exposed by the earlier Pathfinder team. Inside, the Chania Facility was wounded, its primary heat exchanger crippled, its continued existence reliant on the unstable legacy of Project Chimera sleeping fitfully in the rock below. Time had run out for subtlety, for incremental progress.

"Belgrade," Nathan's voice was low, stripped of its usual calm, broadcast only to the core science team. "Status on deliberate Anomaly interface modeling?"

"Fragmentary, Nathan, dangerously so," Belgrade replied, her face pale on the monitor. Ekota stood beside her, looking terrified but resolute. "The recovered Chimera data suggests the Pauli Anomaly responds exponentially to resonant frequencies above a certain threshold. It might act as an energy sink, as hoped or it might cascade, triggering particle decay chains or localized temporal shearing. We lack sufficient data to predict the outcome with certainty."

"Certainty is a luxury we no longer possess," Nathan stated flatly. "Breaching teams are projected to reach Level Beta containment doors within fifteen minutes. Heavy units are mobilizing. Sophia reports increasing EM instability from Level Delta even with the Core offline. The Anomaly is stirring regardless. Our choice is simple: attempt to control the interaction, or be consumed by it when intruders inevitably try to seize this facility." He paused, the weight of command heavy in the silence. "Prepare to initiate controlled resonance induction, targeted thermal dissipation profile. Montane, coordinate field geometry with Sophia's real-time readings from the conduit junction. Sophia, David, secure yourselves and monitor conduit integrity and particle emissions. This is not a deterrent pulse; try to channel the energy/entropy vortex."

A grim understanding settled over the team. This wasn't just defense; it was a radical leap into the unknown, attempting to actively use the dangerous Chimera rift not just as a weapon, but potentially as the key component of the entropy reversal system itself–channeling the immense waste heat into the Anomaly. It was either the breakthrough that justified everything, or their final, catastrophic attempt.

Sophia and David braced themselves within the echoing cavern of Level Delta, near the damaged geothermal conduit junction. Their environmental suits offered protection, but felt woefully inadequate against the forces potentially about to be unleashed. "EM field fluctuations intensifying," Sophia reported, her voice tight. "The rock around the conduit is exhibiting piezo-electric effects–it's under immense stress. We're seeing trace emissions of… Tachyons? Muons? Readings are inconsistent, outside known parameters." The final Chimera logs flashed through her mind "*unidentified, uncontained.*"

"Montane, initiate primary resonance field, low power, tuned to the seventh harmonic as per Chimera cascade signature," Belgrade instructed, her voice strained but clear. Montane's hands moved with precision over the console. The Resonant Field Core spun up, not with the raw power of previous tests, but with a complex, almost musical hum, carefully tuned to probe, to resonate with, the fifty-two-year-old wound in the Earth.

As Montane slowly increased the Core's power, modulating the resonant frequency based on Sophia's real-time feedback from Level Delta, the effects began. Not a sudden blast, but a growing sense of pressure, a building vibration felt throughout the entire facility. Lights flickered erratically. Computer systems struggled to maintain stability against the pervasive, complex energy fields. Ekota fought to keep essential life support and monitoring systems online.

"Anomaly resonance confirmed," Sophia reported, her voice barely audible over the rising hum and suit static. "It's… coupling. Energy draw from the Core is increasing, but reactor temperature is… dropping? No, stabilizing! Thermal energy is being shunted somewhere…into the conduit? Into the rift?"

Belgrade watched the readouts, her eyes wide with a mixture of terror and elation. "It's working! The anomaly is acting as a sink! The resonant field is directing the thermal load into the Pauli Anomaly!" She started inputting commands, refining the field geometry. "Increase power slowly, Montane! Match resonance frequency drift!"

But the Anomaly was not a passive drainpipe. As more energy flowed into it, the strange particle emissions Sophia detected intensified. Temporal micro-fluctuations, previously dismissed as sensor noise, became measurable within Level Delta. David pointed frantically towards the conduit junction. "Sophia, look! The junction housing…it's phasing! Parts of it are becoming momentarily transparent!"

"Spatial instability!" Sophia yelled into the comm. "The energy flux is warping local spacetime near the conduit! Abort! It's unstable!"

Too late. In Level Beta, the Pathfinder Two team reached a heavily reinforced blast door leading towards the suspected core. As their breaching charges detonated, the Anomaly, hyper-stimulated by the resonant field coupling and perhaps the percussive shockwave, reacted violently. The deep hum became a dissonant shriek resonating through the mountain. The blue traceries Sophia had seen earlier erupted from the conduit junction, not as faint sparks, but as blinding arcs of turquoise energy that tore through the rock of Level Delta, vaporizing sections of the cavern wall. The energy cascade propagated upwards. The intrusion team in Level Beta was vaporized, their position on the tactical display vanishing in a burst of static and incoherent energy readings that overloaded Nathan's sensors. The blast doors they were breaching buckled outwards, fused into slag. Throughout the upper levels of the facility, systems failed catastrophically. Lights exploded. Consoles sparked and died. Containment fields flickered and collapsed. In the main control room, emergency lighting kicked in, barely illuminating the chaos. The Resonant Field Core went dark, its connection severed. Comms went dead. Belgrade, Montane, and Ekota were thrown from their stations by the violent tremor that shook the entire facility. Nathan found himself plunged into near silence, his primary displays black, relying on hardened, independent backup systems.

Outside, the effect was less overtly destructive but profoundly bizarre. The Endevor and the intruder vessels registered a massive, complex energy burst from Akrotiri, far exceeding the earlier pulse. But this time, it wasn't just EM interference. Sensors designed to detect gravitational waves registered a powerful, localized distortion. Instruments measuring atmospheric composition detected fleeting spikes of exotic isotopes. High-resolution imagers captured a momentary, shimmering distortion above the peninsula, like the atmosphere itself had warped and buckled before snapping back into place.

Deep within Level Delta, Sophia and David picked themselves up amidst the debris. The cavern was drastically altered. Sections of the wall had vanished, revealing raw, faintly glowing rock. The geothermal conduit junction was a molten ruin. The intense energy discharge had ceased, but the air shimmered with residual heat and strange optical effects. Their environmental suits registered alarming levels of complex radiation and exotic particle contamination. "Nathan? Belgrade? Anyone?" Sophia shouted into her now-static-filled comms. Silence. They were cut off, trapped in the most dangerous part of the facility, possibly irradiated, with the Pauli Anomaly now demonstrably, terrifyingly active.

Hours later, rudimentary systems began flickering back to life within the Chania Facility's upper levels. Nathan re-established comms with the main control room, finding Belgrade, Montane, and Ekota bruised but alive amidst the wreckage. Power was intermittent, life support functioning minimally on emergency backups. The Resonant Field Core was offline, likely damaged. Level Beta was breached and showed signs of intense energy discharge, but no sign of the intrusion team remained. All contact with Sophia and David in Level Delta was lost.

"What... what did we just do?" Ekota asked, his voice trembling, staring at the dead consoles.

Belgrade sat amidst the debris, looking not triumphant, but totally numb. "We didn't just reverse entropy locally. We interfaced with...something else. A consequence of past actions, amplified by our own. The Pauli Anomaly...it's not just a sink. It's a gateway, or a rift, or perhaps both. It channels energy, but it transforms it, releases things we don't understand." She looked towards the flatlined screen where Sophia's vital signs should have been displayed. "The cost..."

JoAnn Solo watched the panicked withdrawal of the Endevor from her hidden position. OPSEC sensors confirmed the massive energy event, the exotic particle signatures, the spatial distortions. She finally broke silence, the message stark: "Containment failed. Pauli Anomaly active, properties unknown. Chania Facility designated Code Omega Contingency. All OPSEC assets withdraw immediately until reassessment complete. No further support authorized for the time being. Maintain monitoring from extreme range. Omega Contingency. Abandonment." JoAnn felt a cold hollowness. They had nurtured this desperate hope, protected it, and now, having touched something truly profound, it had become too dangerous, unleashed beyond control. She transmitted the abort codes to Kim, Reuben, and Ben still awaiting extraction, rerouting them to a deep backup safe zone. The mission was over. The consequences were just beginning.

The world of 2050 continued its slow grind of collapse, famine, and conflict, largely unaware of the specific events beneath Crete. But something fundamental had shifted. Humanity, in its desperate bid to undo the damage of its past, had potentially unlocked a power far greater and more dangerous than global warming itself. The Chania Facility, now a damaged, isolated laboratory haunted by the legacy of two eras of innovative science, stood as a monument to the unknown forces of nature. They had peered beyond entropy, glimpsed the staggering power needed to reshape reality, and found it inextricably bound to forces utterly alien and potentially uncontrollable.

The faint embers of hope scattered across the globe seemed dimmer now, overshadowed by the uncertain dawn of an age where the laws of physics themselves might be rewritten, for better or for terrifyingly worse. The path forward was no longer simply about survival; it was about navigating the terrifying possibilities and responsibilities of wielding power that defied the very laws of the universe. But tomorrow the Sun would come up over the Sea of Crete like always and the Chania Facility team would regroup and try to find another way...